Vibration Assisted Machining

Wiley-ASME Press Series

Advanced Multifunctional Lightweight Aerostructures: Design, Development, and Implementation
Kamran Behdinan and Rasool Moradi-Dastjerdi

Vibration Assisted Machining: Theory, Modelling and Applications
Lu Zheng, Wanqun Chen, Dehong Huo

Two-Phase Heat Transfer
Mirza Mohammed Shah

Computer Vision for Structural Dynamics and Health Monitoring
Dongming Feng, Maria Q Feng

Theory of Solid-Propellant Nonsteady Combustion
Vasily B. Novozhilov, Boris V. Novozhilov

Introduction to Plastics Engineering
Vijay K. Stokes

Fundamentals of Heat Engines: Reciprocating and Gas Turbine Internal Combustion Engines
Jamil Ghojel

Offshore Compliant Platforms: Analysis, Design, and Experimental Studies
Srinivasan Chandrasekaran, R. Nagavinothini

Computer Aided Design and Manufacturing
Zhuming Bi, Xiaoqin Wang

Pumps and Compressors
Marc Borremans

Corrosion and Materials in Hydrocarbon Production: A Compendium of Operational and Engineering Aspects
Bijan Kermani and Don Harrop

Design and Analysis of Centrifugal Compressors
Rene Van den Braembussche

Case Studies in Fluid Mechanics with Sensitivities to Governing Variables
M. Kemal Atesmen

The Monte Carlo Ray-Trace Method in Radiation Heat Transfer and Applied Optics
J. Robert Mahan

Dynamics of Particles and Rigid Bodies: A Self-Learning Approach
Mohammed F. Daqaq

Primer on Engineering Standards, Expanded Textbook Edition
Maan H. Jawad and Owen R. Greulich

Engineering Optimization: Applications, Methods and Analysis
R. Russell Rhinehart

Compact Heat Exchangers: Analysis, Design and Optimization using FEM and CFD Approach
C. Ranganayakulu and Kankanhalli N. Seetharamu

Robust Adaptive Control for Fractional-Order Systems with Disturbance and Saturation
Mou Chen, Shuyi Shao, and Peng Shi

Robot Manipulator Redundancy Resolution
Yunong Zhang and Long Jin

Stress in ASME Pressure Vessels, Boilers, and Nuclear Components
Maan H. Jawad

Combined Cooling, Heating, and Power Systems: Modeling, Optimization, and Operation
Yang Shi, Mingxi Liu, and Fang Fang

Applications of Mathematical Heat Transfer and Fluid Flow Models in Engineering and Medicine
Abram S. Dorfman

Bioprocessing Piping and Equipment Design: A Companion Guide for the ASME BPE Standard
William M. (Bill) Huitt

Nonlinear Regression Modeling for Engineering Applications: Modeling, Model Validation, and Enabling Design of Experiments
R. Russell Rhinehart

Geothermal Heat Pump and Heat Engine Systems: Theory and Practice
Andrew D. Chiasson

Fundamentals of Mechanical Vibrations
Liang-Wu Cai

Introduction to Dynamics and Control in Mechanical Engineering Systems
Cho W.S. To

Vibration Assisted Machining

Theory, Modelling and Applications

Lu Zheng
Newcastle University
Newcastle, UK

Wanqun Chen
Harbin Institute of Technology
Harbin, China

Dehong Huo
Newcastle University
Newcastle, UK

This Work is a co-publication between John Wiley & Sons Ltd and ASME Press.

WILEY

This edition first published 2021

This Work is a co-publication between John Wiley & Sons Ltd and ASME Press.

Registered Offices
John Wiley & Sons, Inc., 111 River Street, Hoboken, NJ 07030, USA
John Wiley & Sons Ltd, The Atrium, Southern Gate, Chichester, West Sussex, PO19 8SQ, UK

Editorial Office
The Atrium, Southern Gate, Chichester, West Sussex, PO19 8SQ, UK

For details of our global editorial offices, customer services, and more information about Wiley products visit us at www.wiley.com.

Wiley also publishes its books in a variety of electronic formats and by print-on-demand. Some content that appears in standard print versions of this book may not be available in other formats.

Library of Congress Cataloging-in-Publication Data

Names: Huo, Dehong, author.
Title: Vibration assisted machining : theory, modelling and applications /
Dehong Huo, Newcastle University, Newcastle, UK, Wanqun Chen, Harbin
Institute of Technology, Harbin, China, Lu Zheng, Newcastle University
Newcastle, UK.
Description: First edition. | Hoboken, NJ : Wiley, 2021. | Series:
Wiley-ASME Press series | Includes bibliographical references.
Identifiers: LCCN 2020027991 (print) | LCCN 2020027992 (ebook) | ISBN
9781119506355 (cloth) | ISBN 9781119506324 (adobe pdf) | ISBN
9781119506362 (epub)
Subjects: LCSH: Machining. | Machine-tools–Vibration. |
Cutting–Vibration. | Machinery, Dynamics of.
Classification: LCC TJ1185 .H87 2021 (print) | LCC TJ1185 (ebook) | DDC
671.3/5–dc23
LC record available at https://lccn.loc.gov/2020027991
LC ebook record available at https://lccn.loc.gov/2020027992

Cover Design: Wiley
Cover Image: © microstock3D/Shutterstock

Set in 9.5/12.5pt STIXTwoText by SPi Global, Chennai, India
Printed and bound by CPI Group (UK) Ltd, Croydon, CR0 4YY

10 9 8 7 6 5 4 3 2 1

Contents

Preface *xi*

1 Introduction to Vibration-Assisted Machining Technology *1*
1.1 Overview of Vibration-Assisted Machining Technology *1*
1.1.1 Background *1*
1.1.2 History and Development of Vibration-Assisted Machining *2*
1.2 Vibration-Assisted Machining Process *3*
1.2.1 Vibration-Assisted Milling *3*
1.2.2 Vibration-Assisted Drilling *3*
1.2.3 Vibration-Assisted Turning *5*
1.2.4 Vibration-Assisted Grinding *5*
1.2.5 Vibration-Assisted Polishing *6*
1.2.6 Other Vibration-Assisted Machining Processes *7*
1.3 Applications and Benefits of Vibration-Assisted Machining *7*
1.3.1 Ductile Mode Cutting of Brittle Materials *7*
1.3.2 Cutting Force Reduction *8*
1.3.3 Burr Suppression *8*
1.3.4 Tool Life Extension *8*
1.3.5 Machining Accuracy and Surface Quality Improvement *9*
1.3.6 Surface Texture Generation *10*
1.4 Future Trend of Vibration-Assisted Machining *10*
References *12*

2 Review of Vibration Systems *17*
2.1 Introduction *17*
2.2 Actuators *18*
2.2.1 Piezoelectric Actuators *18*
2.2.2 Magnetostrictive Actuators *18*
2.3 Transmission Mechanisms *18*
2.4 Drive and Control *19*
2.5 Vibration-Assisted Machining Systems *19*
2.5.1 Resonant Vibration Systems *19*
2.5.1.1 1D System *20*

2.5.1.2 2D and 3D Systems *23*
2.5.2 Nonresonant Vibration System *27*
2.5.2.1 2D System *29*
2.5.2.2 3D Systems *34*
2.6 Future Perspectives *35*
2.7 Concluding Remarks *36*
References *37*

3 Vibration System Design and Implementation *45*
3.1 Introduction *45*
3.2 Resonant Vibration System Design *46*
3.2.1 Composition of the Resonance System and Its Working Principle *46*
3.2.2 Summary of Design Steps *46*
3.2.3 Power Calculation *47*
3.2.3.1 Analysis of Working Length L_{pu} *48*
3.2.3.2 Analysis of Cutting Tool Pulse Force F_p *49*
3.2.3.3 Calculation of Total Required Power *49*
3.2.4 Ultrasonic Transducer Design *49*
3.2.4.1 Piezoelectric Ceramic Selection *49*
3.2.4.2 Calculation of Back Cover Size *51*
3.2.4.3 Variable Cross-Sectional, One-Dimensional Longitudinal Vibration Wave Equation *51*
3.2.4.4 Calculation of Size of Longitudinal Vibration Transducer Structure *53*
3.2.5 Horn Design *53*
3.2.6 Design Optimization *54*
3.3 Nonresonant Vibration System Design *55*
3.3.1 Modeling of Compliant Mechanism *56*
3.3.2 Compliance Modeling of Flexure Hinges Based on the Matrix Method *56*
3.3.3 Compliance Modeling of Flexure Mechanism *59*
3.3.4 Compliance Modeling of the 2 DOF Vibration Stage *61*
3.3.5 Dynamic Analysis of the Vibration Stage *62*
3.3.6 Finite Element Analysis of the Mechanism *63*
3.3.6.1 Structural Optimization *63*
3.3.6.2 Static and Dynamic Performance Analysis *63*
3.3.7 Piezoelectric Actuator Selection *65*
3.3.8 Control System Design *66*
3.3.8.1 Control Program Construction *66*
3.3.9 Hardware Selection *66*
3.3.10 Layout of the Control System *68*
3.4 Concluding Remarks *68*
References *69*
3.A Appendix *70*

4 Kinematics Analysis of Vibration-Assisted Machining *73*
4.1 Introduction *73*
4.2 Kinematics of Vibration-Assisted Turning *74*
4.2.1 TWS in 1D VAM Turning *75*

4.2.2 TWS in 2D VAM Turning *78*
4.3 Kinematics of Vibration-Assisted Milling *80*
4.3.1 Types of TWS in VAMilling *81*
4.3.1.1 Type I *81*
4.3.1.2 Type II *82*
4.3.1.3 Type III *82*
4.3.2 Requirements of TWS *83*
4.3.2.1 Type I Separation Requirements *83*
4.3.2.2 Type II Separation Requirements *85*
4.3.2.3 Type III Separation Requirements *87*
4.4 Finite Element Simulation of Vibration-Assisted Milling *89*
4.5 Conclusion *93*
References *93*

5 Tool Wear and Burr Formation Analysis in Vibration-Assisted Machining *95*
5.1 Introduction *95*
5.2 Tool Wear *95*
5.2.1 Classification of Tool Wear *95*
5.2.2 Wear Mechanism and Influencing Factors *96*
5.2.3 Tool Wear Reduction in Vibration-Assisted Machining *98*
5.2.3.1 Mechanical Wear Suppression in 1D Vibration-Assisted Machining *98*
5.2.3.2 Mechanical Wear Suppression in 2D Vibration-Assisted Machining *101*
5.2.3.3 Thermochemical Wear Suppression in Vibration-Assisted Machining *102*
5.2.3.4 Tool Wear Suppression in Vibration-Assisted Micromachining *106*
5.2.3.5 Effect of Vibration Parameters on Tool Wear *107*
5.3 Burr Formation *108*
5.4 Burr Formation and Classification *109*
5.5 Burr Reduction in Vibration Assisted Machining *109*
5.5.1 Burr Reduction in Vibration-Assisted Micromachining *111*
5.6 Concluding Remarks *113*
5.6.1 Tool Wear *113*
5.6.2 Burr Formation *115*
References *115*

6 Modeling of Cutting Force in Vibration-Assisted Machining *119*
6.1 Introduction *119*
6.2 Elliptical Vibration Cutting *120*
6.2.1 Elliptical Tool Path Dimensions *120*
6.2.2 Analysis and Modeling of EVC Process *120*
6.2.2.1 Analysis and Modeling of Tool Motion *120*
6.2.2.2 Modeling of Chip Geometric Feature *120*
6.2.2.3 Modeling of Transient Cutting Force *124*
6.2.3 Validation of the Proposed Method *126*
6.3 Vibration-Assisted Milling *127*
6.3.1 Tool–Workpiece Separation in Vibration Assisted Milling *128*

6.3.2 Verification of Tool–Workpiece Separation *131*
6.3.3 Cutting Force Modeling of VAMILL *133*
6.3.3.1 Instantaneous Uncut Thickness Model *133*
6.3.3.2 Cutting Force Modeling of VAMILL *136*
6.3.4 Discussion of Simulation Results and Experiments *137*
6.4 Concluding Remarks *143*
References *143*

7 Finite Element Modeling and Analysis of Vibration-Assisted Machining *145*
7.1 Introduction *145*
7.2 Size Effect Mechanism in Vibration-Assisted Micro-milling *147*
7.2.1 FE Model Setup *148*
7.2.2 Simulation Study on Size Effect in Vibration-Assisted Machining *151*
7.3 Materials Removal Mechanism in Vibration-Assisted Machining *152*
7.3.1 Shear Angle *152*
7.3.2 Simulation Study on Chip Formation in Vibration-Assisted Machining *154*
7.3.3 Characteristics of Simulated Cutting Force and von-Mises Stress in Vibration-Assisted Micro-milling *156*
7.4 Burr Control in Vibration-Assisted Milling *158*
7.4.1 Kinematics Analysis *159*
7.4.2 Finite Element Simulation *160*
7.5 Verification of Simulation Models *161*
7.5.1 Tool Wear and Chip Formation *162*
7.5.2 Burr Formation *163*
7.6 Concluding Remarks *164*
References *164*

8 Surface Topography Simulation Technology for Vibration-Assisted Machining *167*
8.1 Introduction *167*
8.2 Surface Generation Modeling in Vibration-Assisted Milling *171*
8.2.1 Cutter Edge Modeling *172*
8.2.2 Kinematics Analysis of Vibration-Assisted Milling *173*
8.2.3 Homogeneous Matrix Transformation *174*
8.2.3.1 Basic Theory of HMT *174*
8.2.3.2 Establishment of HTM in the End Milling Process *174*
8.2.3.3 HMT in VAMILL *176*
8.2.4 Surface Generation *185*
8.2.4.1 Surface Generation Simulation *185*
8.3 Vibration-Assisted Milling Experiments *187*
8.4 Discussion and Analysis *187*
8.4.1 The Influence of the Vibration Parameters on the Surface Wettability *188*
8.4.2 Tool Wear Analysis *189*
8.5 Concluding Remarks *189*
References *189*

Index *193*

Preface

Precision components are increasingly in demand for various engineering industries, such as biomedical engineering, MEMS, electro-optics, aerospace, and communications. However, processing these difficult-to-machine materials efficiently and economically is always a challenging task, which stimulates the development and subsequent application of vibration-assisted machining (VAM) over the past few decades. Vibration-assisted machining employs additional external energy sources to generate high-frequency vibration in the conventional machining process, changing the machining (cutting) mechanism, thus reducing the cutting force and cutting heat and improving the machining quality. The effective implementation of the VAM process depends on a wide range of technical issues, including vibration device design and setup, process parameter optimization, and performance evaluation. The current awareness on VAM technology is incomplete; although ample review/research papers have been published, no single source provides a comprehensive comprehending yet. Therefore, a book is needed to systematically introduce this emerging manufacturing technology as a subject.

The main objective of this book is to address the basics and the latest advances in the VAM technology. The first chapter provides a brief introduction to VAM technology, including VAM process, benefits, and applications, as well as its history and development, so that the reader would have a general understanding of the subject. The second and third chapters aim to present a detailed description of the characteristics and design process for vibration devices. Chapter 2 overviews the current proposed vibration devices in the literature, and the features of each type vibration devices are critically reviewed. Chapter 3 focuses on the implementation and design of vibration devices and the corresponding design procedures are also discussed. Chapters 4 and 5 are dedicated to the effect of vibration and machining parameters on tool path/tool–workpiece separation and the surface topography generation. Chapters 4 and 5 are dedicated to the effect of vibration and machining parameters on tool path/tool–workpiece separation and its influence on the cutting performance. Chapter 4 covers the kinematic analysis of VAM, including the tool–workpiece separation type and the corresponding equations during the processing. Chapter 5 investigates the mechanisms of tool wear and burr generation under different tool–workpiece separation situations. Chapter 6 and 7 investigate VAM process through simulation modelling method. Chapter 6 models the cutting force using both numerical and finite element methods. Finite element modeling and analysis of VAM are detailed in Chapter 7 to deeply understand the cutting mechanism of VAM. The last chapter contains the modeling of surface topography using homogeneous matrix transformation and cutter edge sweeping technology, and the results are verified by the machining experiments.

This book provides state of the art in research and engineering practice in VAM for researchers and engineers in the field of mechanical and manufacturing engineering. This book can be used as a textbook for a final year elective subject on manufacturing engineering, or as an introductory subject on advanced manufacturing methods at the postgraduate level. It can also be used as a textbook for teaching advanced manufacturing technology in general. The book can also serve as a useful reference for manufacturing engineers, production supervisors, tooling engineers, planning and application engineers, as well as machine tool designers.

Some of the research findings in this book have arisen from an EPSRC-funded project "Development of a 3D Vibration Assisted Machining System." The authors gratefully acknowledge the financial support of the Engineering and Physical Sciences Research Council (EP/M020657/1).

The authors wish the readers an enjoyable and fruitful reading through the book.

Newcastle upon Tyne, UK *Lu Zheng, Wanqun Chen and Dehong Huo*
February 2020

1

Introduction to Vibration-Assisted Machining Technology

1.1 Overview of Vibration-Assisted Machining Technology

1.1.1 Background

Precision components are increasingly in demand in various engineering fields such as microelectromechanical systems (MEMS), electro-optics, aerospace, automotive, biomedical engineering, and internet and communication technology (ICT) hardware. In addition to the aims of achieving tight tolerances and high-quality surface finishes, many applications also require the use of hard and brittle materials such as optical glass and technical ceramics owing to their superior physical, mechanical, optical, and electronic properties. However, because of their high hardness and usually low fracture toughness, the processing and fabrication of these hard-to-machine materials have always been challenging. Furthermore, the delicate heat treatment required and composite materials in aeronautic or aerospace alloys have caused similar difficulties for precision machining.

It has been reported that excessive tool wear and fracture damage are the main failure modes during the processing of such materials, leading to low surface quality and machining accuracy. Efforts to optimize a conventional machining process to achieve better cutting performance with these materials have never been stopped, and these optimizations include the cutting parameters, tool materials and geometry, and cutting cooling systems in the past decades [1–6]. Generally, harder materials or wear-resistant coatings are applied, and tool geometry is optimized to prevent tool cracking and to reduce wear on wearable positions such as the flank face [5, 7–10]. Cryogenic coolants are used in the machining process, and their input pressure has been optimized to achieve better cooling performance [2, 4, 11]. However, although cutting performance can be improved, the results are often still unsatisfactory.

Efforts to enhance machining performance have revealed that machining quality can be improved using the high-frequency vibration of the tool or workpiece. Vibration-assisted machining (VAM) was first introduced in the late 1950s and has been applied in various machining processes, including both traditional machining (turning, drilling, grinding, and more recently milling) and nontraditional machining (laser machining, electro-discharge machining, and electrochemical machining), and it is now widely used in the precision manufacturing of components made of various materials. VAM adds external energy to the

Vibration Assisted Machining: Theory, Modelling and Applications,
First Edition. Lu Zheng, Wanqun Chen, and Dehong Huo.

conventional machining process and generate high-frequency, low-amplitude vibration in the tool or workpiece, through which a periodic separation between the uncut workpiece and the tool can be achieved. This can decrease the average machining forces and generate thinner chips, which in turn leads to high processing efficiency, longer tool life, better surface quality and form accuracy, and reduced burr generation [12–17]. Moreover, when hard and brittle materials such as titanium alloy, ceramic, and optical glass are involved, the cutting depth in the ductile regime cutting mode can be increased [18]. As a result, the cutting performance can be improved and unnecessary post-processing can be avoided, which allows the production of components with more complex shape features [14]. Nevertheless, there are still many opportunities for technological improvement, and ample scope exists for better scientific understanding and exploration.

VAM may be classified in two ways. The first classification is according to the dimensions in which vibration occurs: 1D, 2D, or 3D VAM. The other classification is based on the vibration frequency range, for example, in ultrasonic VAM and non-ultrasonic VAM. Ultrasonic VAM is the most common type of VAM. It works at a high vibration frequency (usually above 20 kHz), and a resonance vibration device maintains the desired vibration amplitude. Most of its applications are concentrated in the machining of hard and brittle materials because of the fact that high vibration frequency dramatically improves the cutting performance of difficult-to-machine materials. Meanwhile non-ultrasonic VAM uses a mechanical linkage to transmit power to make the device expand and contract, and this can obtain lower but variable vibration frequencies (usually less than 10 kHz). It is easier to achieve closed-loop control because of the low range of operating frequency, which makes it uniquely advantageous in applications such as the generation of textured surface.

1.1.2 History and Development of Vibration-Assisted Machining

The history of vibration technology in VAM can be traced back to the 1940s. During the period of World War II, the high demand for the electrically controlled four-way spool valves mainly used in the control of aircraft and gunnery circuits stimulated the development of servo valve technology [19]. Because of their wide frequency response and high flow capacity, electrohydraulic vibrators were successfully developed and applied in VAM in the 1960s with positive effects in enhanced processing quality and efficiency [20]. With the further development of technology, electromagnetic vibrators featuring higher accuracy and a wide range of frequency and amplitude generation were developed based on electromagnetic technology, and these were successfully applied to various VAM processes [21]. The need for complex hydraulic lines was eliminated, and greater tolerance for the application environment was allowed, which also leads to smaller devices. As a result, a transmission line or connecting body can be attached to the vibrator to achieve a wide range of vibration frequencies and amplitude adjustments [22]. In the 1980s, the maturity of piezoelectric transducer (PZT) piezoelectric ceramic technology had brought a new choice for the vibrator. A piezoelectric ceramic stack could be sandwiched under compressive strain between metal plates, and this has advantages including compactness, high precision and resolution, high frequency response, and large output force [23]. Various shapes of piezoelectric ceramic elements can be used to make different types of vibration actuators, which indicate that the limitations of traditional vibrators were overcome and the application of VAM technology for precision machining was broadened. In addition, it helped in the development of multidimensional VAM equipment. Elliptical VAM has received extensive attention since it was first proposed in the 1990s. Although this process has many advantages compared to its 1D counterpart in terms of reductions in cutting force and prolongation of tool life, it requires higher performance in the vibrator, producing a more accurate tool tip trajectory [24–28]. Piezoelectric actuators with high sensitivity can fulfill the requirements of vibration devices and promote the development of elliptical VAM technology.

1.2 Vibration-Assisted Machining Process

This section briefly introduces commonly used VAM processes, including milling, drilling, turning, grinding, and polishing. Different vibration device layouts are required to implement these vibration-assisted processes and to achieve advantages over the corresponding conventional machining processes.

1.2.1 Vibration-Assisted Milling

Milling is one of the most common machining processes and is capable of fabricating parts with complex 3D geometry. However, uncontrollable vibration problems during the cutting process are quite serious and can affect processing stability, especially in the micro-milling process, leading to excessive tolerance, increased surface roughness, and higher cost. Vibration-assisted milling is a processing method that combines the external excitation of periodic vibrations with the relative motion of the milling tool or workpiece to obtain better cutting performance. In addition to the same advantages as other VAM processes, complex surface microstructures can also be obtained because of the combination of a unique tool path and external vibration. Currently, the application of vibration-assisted milling mainly focuses on the one-dimensional direction. The vibration may be applied in the feed direction, cross-feed direction, or axial direction, and tool rotational vibrations may also be applied [14]. Little research has been carried out on 2D vibration-assisted milling because of the difficulty of developing two-dimensional vibration platforms (motion coupling and control difficulty), and the vibration mode of these 2D vibration devices mainly involves elliptical vibration and longitudinal torsional vibration.

1.2.2 Vibration-Assisted Drilling

Problems such as large axial forces and poor surface quality are found in the process of drilling the hard and brittle materials. Vibration-assisted drilling technology combines the VAM mechanism with the traditional drilling process, and this can achieve more efficient drilling, especially for small bore diameters and deep holes. Compared with conventional drilling, the interaction between the tool and the workpiece is changed, and the drilling tool edge cutting conditions are improved. Vibration-assisted drilling has found applications in the high-efficiency and high-quality machining of various parts with difficult-to-machine holes [29]. Its main merits are as follows:

(1) *Reductions in drilling power and drilling torque.* The vibration changes the interaction between the drill tool and the workpiece, and the cutting process changes from continuous cutting to intermittent cutting, leading to lower tool axial force. In addition, the friction factor between the tool and the workpiece/chips is reduced because of the pulse torque formed by the vibration. As a result, drilling torque is reduced [30, 31].
(2) *Improvement in chip breaking and removal performance.* The chip breaking mechanism is quite different when vibration is added. Fragmented chips can be obtained under certain vibration and machining parameters. Chip removal performance is much better compared with the continuous chips produced in conventional drilling [32].
(3) *Improvement in the surface quality of the walls of the drilled holes.* In the vibration-assisted drilling process, the reciprocal pressing action of the cutting edge on the inner hole surface is beneficial in reducing surface roughness. Moreover, the improved chip breaking performance also leads to smoother chip removal, which reduces the scratching of the drilled hole surface by chips and the surface roughness [33, 34].
(4) *Improvement in tool life.* The intermittent cutting improves the drilling tool's cooling conditions, leading to lower cutting temperature and relieving the built-up edge and tool chipping effects. As a result, longer tool life can be obtained [35, 36].

As shown in Figure 1.1, according to the direction of vibration, vibration-assisted drilling can be divided into axial, torsional, and axial–torsional composite vibration drilling. The vibration direction in axial vibration drilling is consistent with the direction of the drilling tool axis, while in torsional vibration drilling, it is consistent with the direction of the drilling tool's rotation. Axial torsional composite vibration drilling combines the previous two types.

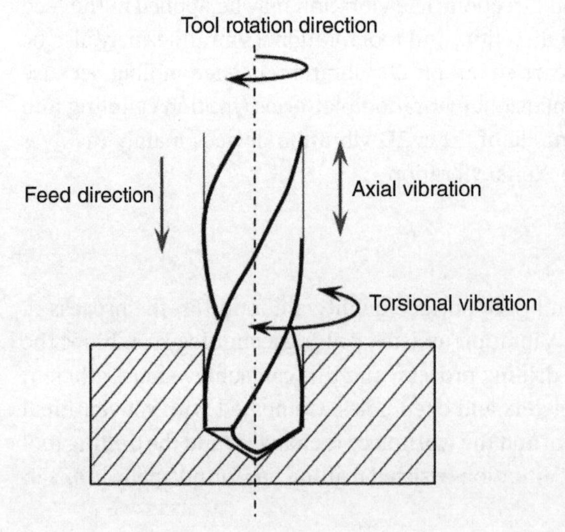

Figure 1.1 Schematic of vibration-assisted drilling.

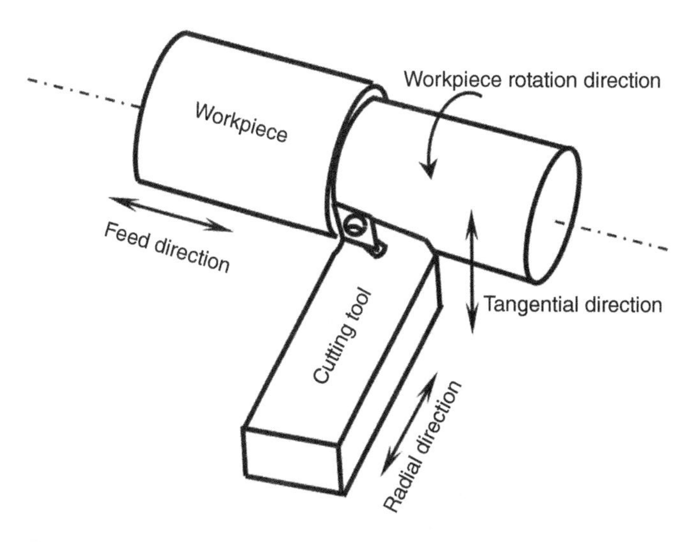

Figure 1.2 Schematic of vibration-assisted turning.

1.2.3 Vibration-Assisted Turning

Turning is a widely used machining method because of its high processing quality, metal removal rate, and productivity and efficient equipment utilization. However, drawbacks such as large cutting forces, difficulties in chip removal, and serious tool wear can cause serious processing problems, such as low machined quality and efficiency and high cost. Vibration-assisted turning provides a new method for the efficient and high-quality machining of difficult materials. As shown in Figure 1.2, vibration is applied to the turning tool mainly in the radial, tangential, and feed directions. Multidimensional vibration-assisted turning is generally referred as elliptical vibration-assisted turning, where two of the above three vibration directions are chosen and applied to the turning tool. One-dimensional vibration-assisted turning represents a large proportion of methods of vibration-assisted turning proposed so far. Most apply vibration in the feed direction, and experimental results have proven that this has a significant influence in reducing cutting forces, cutting temperature, and improving the quality of processing. Currently, only a few studies have applied vibration in the other directions, and the effects and cutting mechanisms of involved in material processing need further research.

1.2.4 Vibration-Assisted Grinding

Compared with other machining processes, grinding is increasingly used in the field of ultraprecision/precision machining because of its better machining accuracy and surface roughness. However, processing material with grinding wheels is a complex and stochastic process, where the ground surface may become damaged and low wheel life is caused by the high grinding forces and high surface cutting temperature (as the grinding wheel instantaneous temperature can reach 1000 °C). Vibration-assisted grinding process applies vibration to the grinding wheel or workpiece during the grinding so as to improve the material removal performance. The vibration can be applied in the tangential, radial, or

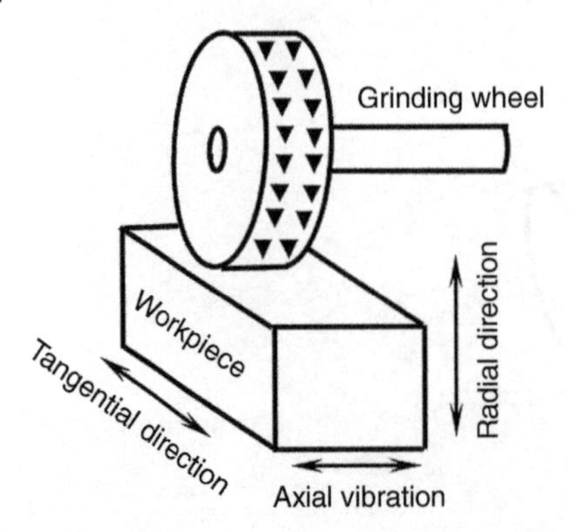

Figure 1.3 Schematic of vibration-assisted grinding.

axial direction along the grinding wheel, as is shown in Figure 1.3. Vibration-assisted grinding in the tangential and radial directions is similar to intermittent grinding, and tool–workpiece separation can be obtained during the machining process. Although vibration-assisted grinding in the axial direction involves a continuous grinding process, the machining process is quite different in conventional grinding and features separation, impact and reciprocating ironing characteristics, and lubrication effects, which can reduce grinding wheel blockage, cutting forces, workpiece residual stress, and machined surface burn. As a result, better processing performance and longer tool life can be obtained. In addition, it can also effectively reduce the chipping of hard and brittle workpiece materials and surface or subsurface cracking as well as machined surface quality [37–39]. Although similar to the mechanism of other VAM processes, the randomness of the size, shape, and distribution of abrasive grains on the grinding wheel surface and the complexity of the grinding motion bring great challenges to the study of the mechanisms involved in the vibration-assisted grinding.

1.2.5 Vibration-Assisted Polishing

At present, various miniature optical lenses are generally fabricated by precision injection molding with silicon carbide or tungsten carbides molds, and these molds usually require polishing to achieve optical grade surface quality. However, small mold sizes and increasingly high precision requirements make the polishing process challenging. In the conventional polishing process, the high-speed rotation of soft polishing tools such as wool, rubber, and asphalt polishing heads are often used to process the workpiece surface. However, when the surface has complex curved shapes and a small curvature radius, the complicated polishing mechanism and uncontrollable polishing forces severely limit the processing results. Vibration-assisted polishing can overcome some of the shortcomings in conventional polishing. Using this method, the polishing head does not need to be rotated at such high speeds, which helps in ensuring constant polishing force during the polishing process, which can also be used for smaller size molds. Current research shows that vibration-assisted polishing can improve the surface roughness of the polished workpiece and the surface accuracy while achieving high polishing efficiency [40–42].

1.2.6 Other Vibration-Assisted Machining Processes

With the advantages of VAM gradually being demonstrated, more machining processes are being added to the VAM family. Two examples of the newly developed VAM processes are vibration-assisted boring and vibration-assisted electrical discharge machining. In order to solve the difficult machining problem of complex deep hole parts with high length-to-diameter ratios (>20) such as aeroengine fuel nozzles, vibration-assisted boring has been developed. Compared with the conventional boring process, the tool can be prevented from colliding with the machined surface during the separation stage, the plastic critical cutting depth of the brittle material is increased, and cutting edge cracking and cutting tool flank face reverse bulges are avoided [43, 44]. Its separation and reversal characteristics can greatly reduce the radial thrust force and effectively improve the absolute stability of the cutting stiffness. As a result, the machined surface quality is improved and cutting flutter can be suppressed. Vibration-assisted electrical discharge machining has also been successfully applied in processing micro-hole parts made of hard and brittle materials [45–47]. In conventional electrical discharge machining, the discharge gap between the tool and the workpiece is usually only a few micrometers to several tens of micrometers and easily causes the deterioration due to the slag discharge effect and local concentration of processing debris, causing abnormal discharges and reducing processing efficiency. Compared with the process, vibration-assisted electrical discharge machining has better processing efficiency, and the results show that the slag removal effect is ameliorated and electrode wear reduced.

1.3 Applications and Benefits of Vibration-Assisted Machining

1.3.1 Ductile Mode Cutting of Brittle Materials

When the cutting depth is less than a certain critical value (the critical cutting depth) in the processing of brittle materials, the cutting process will be transformed from brittle cutting mode into ductile cutting mode. This removes the workpiece materials by plastic flow instead of brittle fractures, leading to a crack-free surface. In ductile cutting mode, the critical cutting depth can be defined as the cutting depth at which a crack appears on the machined surface. If the undeformed chip thickness is less than the critical cutting depth, brittle cutting can be reduced in conventional cutting and a better surface finish can be obtained. However, in the actual processing of brittle materials, their critical cutting depth is usually in the range of microns or submicron, which reduces the processing efficiency and increases the manufacturing time. VAM is an effective method used to increase the critical cutting depth in ductile cutting mode and to improve the economics and feasibility of the processing of brittle materials. It has been reported that smaller cutting forces can reduce microcrack propagation on the surface of the brittle parts and can increase the critical cutting depth for brittle materials under ductile cutting mode. In addition, a large enough plastic yielding force, but not large enough to cause material rupturing, is also a necessary condition for the ductile cutting of brittle workpieces. Therefore, it is feasible to increase the brittle materials critical cutting depth within a reasonable stress range by using VAM [48–50].

1.3.2 Cutting Force Reduction

A large number of cutting experiments and finite element analysis show that under the same cutting conditions, the average cutting force of VAM is significantly lower than in the traditional cutting process, and the cutting force in 2D VAM process is less than that in 1D. Although the instantaneous peak cutting force of 1D VAM is close to the steady-state cutting force in conventional machining, a lower average cutting force can be obtained because of the periodic contact between the tool and workpiece during cutting [51–53]. In 2D VAM process, the shape of chips and the interaction between them and the tool rake face are quite different from 1D VAM because of the elliptical cutting tool trajectory, which leads to lower average cutting force and reduced instantaneous peak cutting force. The cutting force reduction is manifested in the following ways:

(1) The chip thickness in the 2D VAM can be reduced because of the continuous overlapping of elliptical tool paths. As a result, the cutting forces in different directions can be reduced.
(2) Under certain conditions such as circular or narrowly elliptical tool paths, the cutting tool moves faster than the chip flow speed, causing reverse friction between the tool and the workpiece, and the back cutting force can be reduced or even reversed.
(3) The periodic contact between the cutting tool and workpiece improves the lubrication conditions during the cutting process and facilitates the dissipation of heat from the tool, resulting in a reduction in cutting force.

1.3.3 Burr Suppression

Burr formation, similar to chip generation, is a common and undesirable phenomenon in the machining process and is one of the most important criteria in the evaluation of the machined surface. VAM can effectively suppress burr formation during processing, and some researchers have proposed that burr height can be reduced up to 80% compared with conventional machining [54–56]. Figure 1.4 shows examples of burr reduction in VAM. Almost no burrs can be found on the machined surface. This phenomenon is mainly due to the reduced cutting force, which leads to lower transient compressive stress and yield stress in the cutting deformation area. In addition, unique tool trajectories (such as elliptical trajectories) can result in discrete small pieces of chips. As a result, burr formation can be suppressed.

1.3.4 Tool Life Extension

Machining processes are inherently involved in tool wear, which is usually evaluated in terms of average cutting force, machined surface roughness, and cumulative cutting length. It has an important impact on surface quality and machining costs. VAM can effectively

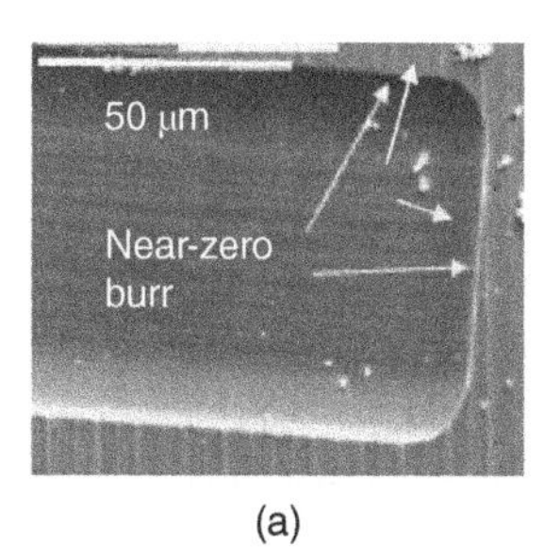

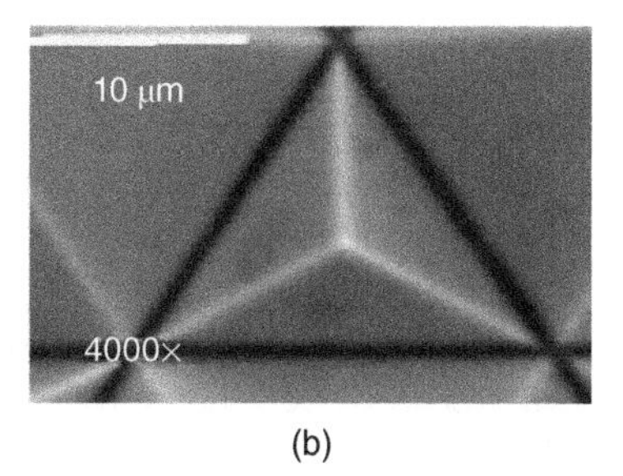

(a) (b)

Figure 1.4 SEM images of burr-free structures made using 2D VAM. Single-crystal diamond tool in hard-plated copper. (a) Microchannel, 1.5 μm deep, and (b) a 8 μm tall regular trihedron made using a dead-sharp tool with a 70° nose angle. Source: Brehl and Dow [14]. © 2008, Elsevier.

improve cutting tool life, especially in the processing of hard materials. Unlike the irregular wear caused by traditional machining tools, the tool wear in VAM is smooth and inclined. At lower spindle speeds, due to the lower cutting temperatures, the dominant wear mechanism is abrasive wear. Because of the mechanical and impact contact between the workpiece and tool flank surface in VAM, tool life is less than that in the conventional process. At higher cutting speeds, temperature-activated wear mechanisms occur, such as diffusion, chemical wear, and thermal wear. On the other hand, because of the intermittent separation of the workpiece and tool, the temperature in the cutting zone in VAM is lower than that in conventional process, which tends to increase the tool life. Another reason for reducing the temperature in VAM is the change in friction coefficient from semi-static to dynamic, which results in a reduced friction coefficient in the process and a change in the chip formation mechanism. As the cutting speed increases, there is an increase in the degree of tool–workpiece engagement per tool revolution. As a result, the effect of vibration on the machining process decreases, and the cutting forces in VAM and conventional milling processes become closer to each other. A detailed analysis on how VAM enhances tool life is provided in Chapter 5.

1.3.5 Machining Accuracy and Surface Quality Improvement

Compared with the conventional machining process, VAM can greatly improve the machining accuracy and surface quality, and the improvements vary depending on the tool and workpiece materials, vibration conditions (vibration amplitude, vibration frequency, and vibration dimensions), tool parameters, and processing parameters such as feed rate, spindle speed, and cutting depth. If the processing parameters are unchanged, the surface roughness in 1D and 2D VAM can be reduced by approximately 40% and 85%, respectively [14]. There are many reasons for this. On the one hand, lower cutting forces can enhance the stability of the cutting process, which reduces tool run-out in the cutting depth direction and generates smaller chips. On the other hand, VAM can reduce cutting tool wear and effectively avoid damage caused to the machined surface by worn tools. The tool's self-excited vibration is replaced with regular sine or cos vibration, which reduces the residual height of the unremoved material. As a result, a better machined surface quality can be obtained.

1.3.6 Surface Texture Generation

Engineered textured surfaces have the characteristics of regular textural structures and high aspect ratio, enabling the component surface to serve specific functions such as reducing adhesion friction, improving lubricity, increasing wear resistance, changing hydrophilic performance, and enhancing optical properties. Etching methods are commonly used to produce high precision surface microstructures, but these are costly and time-consuming. As a more flexible method, it has been proven that VAM in either a single direction or two directions can form certain surface textures depending on the cutting edge geometry and kinematics. Currently, the proposed surface textures mainly include a squamous, micro-dimple pattern and micro-convex pattern types, and their size ranges from a few microns to tens of microns, as shown in Figure 1.5. There is an emerging trend to obtain certain surface performance using VAM. For example, the size of the surface texture features can be controlled by changing the vibration and processing parameters, leading to variable surface wettability (Figure 1.5) [12]. The process can also be used to create microchannels for the microfluidic control of the fluid flow, to name a few. A detailed analysis on how VAM produces surface texture is provided in Chapter 8.

1.4 Future Trend of Vibration-Assisted Machining

With the development of processing technology, the application of VAM is becoming increasingly widespread, and research into VAM is becoming more and more intensive, mainly in the following main aspects:

(1) *Development and adoption of new tool materials.* The proportion of difficult-to-machine materials in modern products is increasing, as well as a higher processing quality of these parts. In order to achieve better cutting performance, in addition to the optimization of tool geometry parameters, more attention has been focused on the development and application of tool materials in VAM, and main research focus is on natural and synthetic diamond and ultrafine grained carbide materials.
(2) *Ultra-high-frequency vibration-assisted machining.* Ultra-high-frequency VAM will continue to be a research focus in VAM in the future. Recent research indicates that the possibility of grinding wheel ablation can be effectively reduced by adding high-frequency vibration, which also improves the grinding wheel's life and the surface quality of the workpiece. In recent years, research into ultra-high-frequency vibration equipment has made it possible to reach a maximum vibration frequency of 100 kHz, and at the same time, its processing performance for brittle and hard materials has also been significantly improved.
(3) *Precision/ultraprecision application.* It has been reported that the dimensional and geometric accuracy and wear resistance as well as corrosion resistance of the workpiece can be improved dramatically when low-frequency vibration is applied. However, only high-frequency VAM, such as ultrasonic VAM, can currently achieve a precision machining process. For example, a surface roughness of Ra 0.02–0.04 μm can be obtained by vibration-assisted honing, and surface quality improves by an order of magnitude in the ultrasonic vibration extrusion process compared to conventional

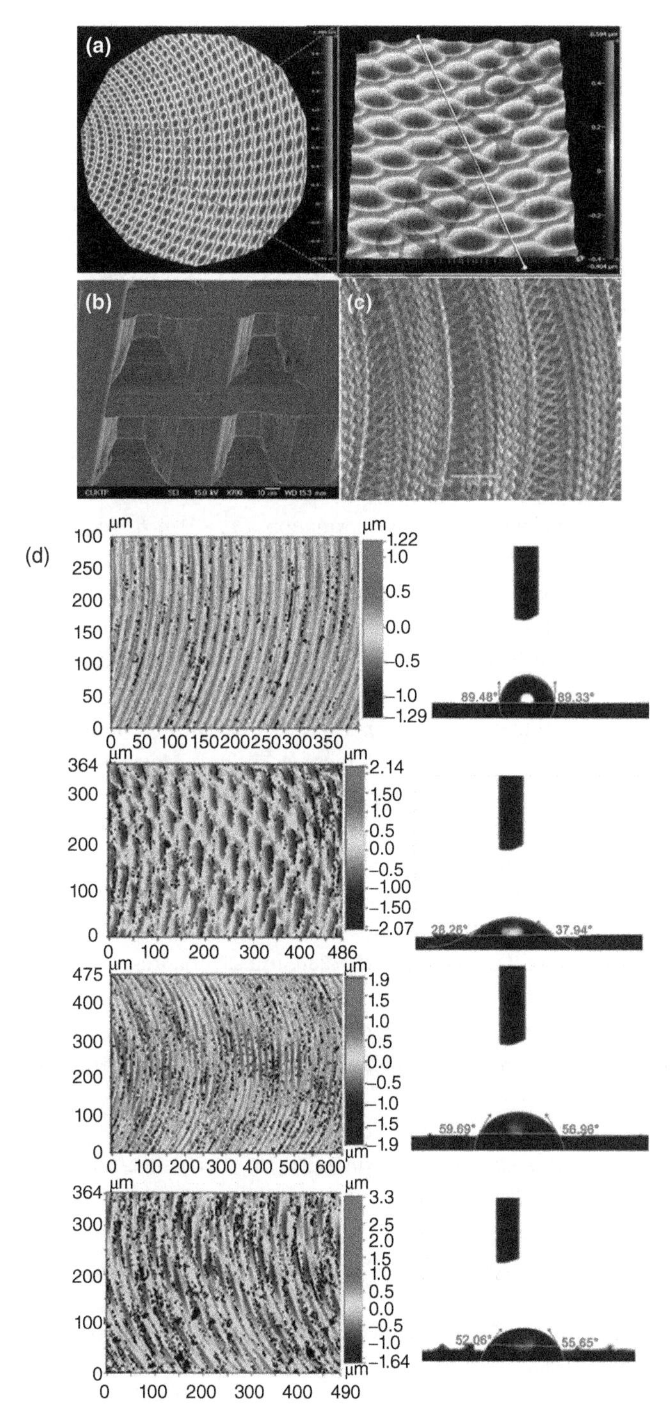

Figure 1.5 Surface texture produced by vibration-assisted machining: (a) micro-dimple patterns. Source: Lin et al. [57]. © 2017, IOP Publishing Ltd, (b) micro-convex patterns. Source: Kim and Loh [58]. © 2010, Springer Nature, (c) squamous patterns. Source: Tao et al. [59]. © 2017, Taylor & Francis Group, and (d) surface wettability variation with different surface textures. Source: Chen et al. [12].

extrusion. Ultrasonic vibration machining can not only guarantee the quality of ultraprecision machining but also allows for higher cutting rates, leading to higher productivity.

(4) *In-depth study of vibration-assisted machining mechanism*. Although the cutting mechanism of VAM has been investigated by several researchers, it is still not fully understood. Current and future research on VAM will focus on several areas, including the effect of the separation and non-separation of the workpiece and cutting tool on chip formation, mechanical analysis of the interaction between the cutting tool and workpiece, microscopic studies, and mathematical descriptions of VAM mechanisms, to name a few.

References

1 Sutter, G. and List, G. (2013). Very high speed cutting of Ti–6Al–4V titanium alloy – change in morphology and mechanism of chip formation. *Int. J. Mach. Tools Manuf.* 66: 37–43. https://doi.org/10.1016/j.ijmachtools.2012.11.004.

2 Da Silva, R.B., MacHado, Á.R., Ezugwu, E.O. et al. (2013). Tool life and wear mechanisms in high speed machining of Ti–6Al–4V alloy with PCD tools under various coolant pressures. *J. Mater. Process. Technol.* 213: 1459–1464. https://doi.org/10.1016/j.jmatprotec.2013.03.008.

3 Sharman, A.R.C., Hughes, J.I., and Ridgway, K. (2015). The effect of tool nose radius on surface integrity and residual stresses when turning Inconel 718™. *J. Mater. Process. Technol.* 216: 123–132. https://doi.org/10.1016/j.jmatprotec.2014.09.002.

4 Sadik, M.I., Isakson, S., Malakizadi, A., and Nyborg, L. (2016). Influence of coolant flow rate on tool life and wear development in cryogenic and wet milling of Ti–6Al–4V. *Procedia CIRP* 46: 91–94. https://doi.org/10.1016/j.procir.2016.02.014.

5 Ulutan, D. and Ozel, T. (2011). Machining induced surface integrity in titanium and nickel alloys: a review. *Int. J. Mach. Tools Manuf.* 51: 250–280. https://doi.org/10.1016/j.ijmachtools.2010.11.003.

6 Ezugwu, E.O., Bonney, J., and Yamane, Y. (2003). An overview of the machinability of aeroengine alloys. *J. Mater. Process. Technol.* 134: 233–253. https://doi.org/10.1016/S0924-0136(02)01042-7.

7 Basturk, S., Senbabaoglu, F., Islam, C. et al. (2010). Titanium machining with new plasma boronized cutting tools. *CIRP Ann. Manuf. Technol.* 59: 101–104. https://doi.org/10.1016/j.cirp.2010.03.095.

8 Ribeiro, M.V., Moreira, M.R., and Ferreira, J.R. (2003). Optimization of titanium alloy (6Al–4V) machining. *J. Mater. Process. Technol.*: 143, 458–144, 463. https://doi.org/10.1016/S0924-0136(03)00457-6.

9 Hatt, O., Crawforth, P., and Jackson, M. (2017). On the mechanism of tool crater wear during titanium alloy machining. *Wear* 374–375: 15–20. https://doi.org/10.1016/j.wear.2016.12.036.

10 Jawaid, A., Sharif, S., and Koksal, S. (2000). Evaluation of wear mechanisms of coated carbide tools when face milling titanium alloy. *J. Mater. Process. Technol.* 99: 266–274. https://doi.org/10.1016/S0924-0136(99)00438-0.

11 MacHai, C. and Biermann, D. (2011). Machining of β-titanium-alloy Ti–10V–2Fe–3Al under cryogenic conditions: cooling with carbon dioxide snow. *J. Mater. Process. Technol.* 211: 1175–1183. https://doi.org/10.1016/j.jmatprotec.2011.01.022.

12 Chen, W., Zheng, L., and Huo, D. (2018). Surface texture formation by non-resonant vibration assisted micro milling. *J. Micromech. Microeng.* 28: 025006. https://doi.org/10.1088/1361-6439/aaa06f.

13 Janghorbanian, J., Razfar, M.R., and Zarchi, M.M.A. (2013). Effect of cutting speed on tool life in ultrasonic-assisted milling process. *Proc. Inst. Mech. Eng. Part B J. Eng. Manuf.* 227: 1157–1164. https://doi.org/10.1177/0954405413483722.

14 Brehl, D.E. and Dow, T.A. (2008). Review of vibration-assisted machining. *Precis. Eng.* 32: 153–172. https://doi.org/10.1016/j.precisioneng.2007.08.003.

15 Lian, H., Guo, Z., Huang, Z. et al. (2013). Experimental research of Al6061 on ultrasonic vibration assisted micro-milling. *Procedia CIRP*: 561–564. https://doi.org/10.1016/j.procir.2013.03.056.

16 Shen, X.H., Zhang, J.H., Li, H. et al. (2012). Ultrasonic vibration-assisted milling of aluminum alloy. *Int. J. Adv. Manuf. Technol.* 63: 41–49. https://doi.org/10.1007/s00170-011-3882-5.

17 Chern, G.L. and Chang, Y.C. (2006). Using two-dimensional vibration cutting for micro-milling. *Int. J. Mach. Tools Manuf.* 46: 659–666. https://doi.org/10.1016/j.ijmachtools.2005.07.006.

18 Zheng, L., Chen, W., and Huo, D. (2018). Experimental investigation on burr formation in vibration-assisted micro-milling of Ti–6Al–4V. *Proc. Inst. Mech. Eng. Part C J. Mech. Eng. Sci.*: 095440621879236. https://doi.org/10.1177/0954406218792360.

19 Ashley, B.C. and Millst, B. (1966). Frequency response of an electro-hydraulic vibrator with inertial load. *J. Mech. Eng. Sci.* 8: 27–35.

20 Skelton, R.C. (1969). Effect of ultrasonic vibration on the turning process. *Int. J. Mach. Tool Des. Res.* 9: 363–374.

21 Lenkiewicz, W. (1969). The sliding friction process – effect of external vibrations. *Wear* 13: 99–108.

22 Balamuth, L. (1964). Recent developments in ultrasonic metalworking processes. Paper presented at SAE/ASME Air Transport and Space Meeting, New York (27–30 April 1964).

23 Xu, C., Akiyama, M., Nonaka, K., and Watanabe, T. (1998). Electrical power generation characteristics of PZT piezoelectric ceramics. *IEEE Trans. Ultrason. Ferroelectr. Freq. Control* 45: 1065–1070.

24 Kumar, M.N., Kanmani Subbu, S., Vamsi Krishna, P., and Venugopal, A. (2014). Vibration assisted conventional and advanced machining: a review. *Procedia Eng.* 97: 1577–1586. https://doi.org/10.1016/j.proeng.2014.12.441.

25 Xu, W.X. and Zhang, L.C. (2015). Ultrasonic vibration-assisted machining: principle, design and application. *Adv. Manuf.* 3: 173–192. https://doi.org/10.1007/s40436-015-0115-4.

26 Shamoto, E. and Moriwaki, T. (1994). Study on elliptical vibration cutting. *CIRP Ann. Manuf. Technol.* 43: 35–38. https://doi.org/10.1016/S0007-8506(07)62158-1.

27 Negishi, N. (2003). *Elliptical Vibration Assisted Machining with Single Crystal Diamond Tools*. North Carolina State University.

28 Shamoto, E., Suzuki, N., and Hino, R. (2008). Analysis of 3D elliptical vibration cutting with thin shear plane model. *CIRP Ann. Manuf. Technol.* 57: 57–60. https://doi.org/10.1016/j.cirp.2008.03.073.

29 Baghlani, V., Mehbudi, P., Akbari, J. et al. (2016). An optimization technique on ultrasonic and cutting parameters for drilling and deep drilling of nickel-based high-strength Inconel 738LC superalloy with deeper and higher hole quality. *Int. J. Adv. Manuf. Technol.* https://doi.org/10.1007/s00170-015-7414-6.

30 Ding, K., Fu, Y., Su, H. et al. (2014). Experimental studies on drilling tool load and machining quality of C/SiC composites in rotary ultrasonic machining. *J. Mater. Process. Technol.* https://doi.org/10.1016/j.jmatprotec.2014.06.015.

31 Alam, K., Mitrofanov, A.V., and Silberschmidt, V.V. (2011). Experimental investigations of forces and torque in conventional and ultrasonically-assisted drilling of cortical bone. *Med. Eng. Phys.* 33: 234–239. https://doi.org/10.1016/j.medengphy.2010.10.003.

32 Chen, S., Zou, P., Tian, Y. et al. (2019). Study on modal analysis and chip breaking mechanism of Inconel 718 by ultrasonic vibration-assisted drilling. *Int. J. Adv. Manuf. Technol.* https://doi.org/10.1007/s00170-019-04155-6.

33 Hsu, I. and Tsao, C.C. (2009). Study on the effect of frequency tracing in ultrasonic-assisted drilling of titanium alloy. *Int. J. Adv. Manuf. Technol.* https://doi.org/10.1007/s00170-008-1696-x.

34 Dvivedi, A. and Kumar, P. (2007). Surface quality evaluation in ultrasonic drilling through the Taguchi technique. *Int. J. Adv. Manuf. Technol.* https://doi.org/10.1007/s00170-006-0586-3.

35 Pecat, O. and Brinksmeier, E. (2014). Tool wear analyses in low frequency vibration assisted drilling of CFRP/Ti6Al4V stack material. *Procedia CIRP* 14: 142–147. https://doi.org/10.1016/j.procir.2014.03.050.

36 Barani, A., Amini, S., Paktinat, H., and Fadaei Tehrani, A. (2014). Built-up edge investigation in vibration drilling of Al2024-T6. *Ultrasonics* https://doi.org/10.1016/j.ultras.2014.01.003.

37 Nik, M.G., Movahhedy, M.R., and Akbari, J. (2012). Ultrasonic-assisted grinding of Ti6Al4V alloy. *Procedia CIRP* https://doi.org/10.1016/j.procir.2012.04.063.

38 Shen, J.Y., Wang, J.Q., Jiang, B., and Xu, X.P. (2015). Study on wear of diamond wheel in ultrasonic vibration-assisted grinding ceramic. *Wear* https://doi.org/10.1016/j.wear.2015.02.047.

39 Chen, J.B., Fang, Q.H., Wang, C.C. et al. (2016). Theoretical study on brittle–ductile transition behavior in elliptical ultrasonic assisted grinding of hard brittle materials. *Precis. Eng.* https://doi.org/10.1016/j.precisioneng.2016.04.005.

40 Shiou, F.J. and Ciou, H.S. (2008). Ultra-precision surface finish of the hardened stainless mold steel using vibration-assisted ball polishing process. *Int. J. Mach. Tools Manuf.* 48: 721–732. https://doi.org/10.1016/j.ijmachtools.2008.01.001.

41 Suzuki, H., Moriwaki, T., Okino, T., and Ando, Y. (2006). Development of ultrasonic vibration assisted polishing machine for micro aspheric die and mold. *CIRP Ann. Manuf. Technol.* https://doi.org/10.1016/S0007-8506(07)60441-7.

42 Yin, S. and Shinmura, T. (2004). A comparative study: polishing characteristics and its mechanisms of three vibration modes in vibration-assisted magnetic abrasive polishing. *Int. J. Mach. Tools Manuf.* 44: 383–390. https://doi.org/10.1016/j.ijmachtools.2003.10.002.

43 Moraru, G.F. (2008). Nonlinear dynamics in drilling and boring operations assisted by low frequency vibration. *2007 Proceedings of the ASME International Design Engineering Technical Conferences and Computers and Information in Engineering Conference, DETC2007*. 4–7 September, 2007, Las Vegas, NV, ASME. doi:https://doi.org/10.1115/DETC2007-35043.

44 Zhang, X., Sui, H., Zhang, D., and Wu, R. (2017). The improvement of deep-hole boring machining quality assisted with ultrasonic vibration. *Jixie Gongcheng Xuebao/Journal Mech. Eng.* https://doi.org/10.3901/JME.2017.19.143.

45 Shabgard, M.R., Badamchizadeh, M.A., Ranjbary, G., and Amini, K. (2013). Fuzzy approach to select machining parameters in electrical discharge machining (EDM) and ultrasonic-assisted EDM processes. *J. Manuf. Syst.* https://doi.org/10.1016/j.jmsy.2012.09.002.

46 Xu, M.G., Zhang, J.H., Li, Y. et al. (2009). Material removal mechanisms of cemented carbides machined by ultrasonic vibration assisted EDM in gas medium. *J. Mater. Process. Technol.* https://doi.org/10.1016/j.jmatprotec.2008.04.031.

47 Uhlmann, E. and Domingos, D.C. (2016). Investigations on vibration-assisted EDM-machining of seal slots in high-temperature resistant materials for turbine components – part II. *Procedia CIRP* https://doi.org/10.1016/j.procir.2016.02.179.

48 Zhang, J., Suzuki, N., Wang, Y., and Shamoto, E. (2014). Fundamental investigation of ultra-precision ductile machining of tungsten carbide by applying elliptical vibration cutting with single crystal diamond. *J. Mater. Process. Technol.* https://doi.org/10.1016/j.jmatprotec.2014.05.024.

49 Du Kim, J. and Choi, I.H. (1997). Micro surface phenomenon of ductile cutting in the ultrasonic vibration cutting of optical plastics. *J. Mater. Process. Technol.* https://doi.org/10.1016/S0924-0136(96)02546-0.

50 Zhou, M., Wang, X.J., Ngoi, B.K.A., and Gan, J.G.K. (2002). Brittle-ductile transition in the diamond cutting of glasses with the aid of ultrasonic vibration. *J. Mater. Process. Technol.* 121: 243–251. https://doi.org/10.1016/S0924-0136(01)01262-6.

51 Zhou, M., Eow, Y.T., Ngoi, B.K.A., and Lim, E.N. (2003). Vibration-assisted precision machining of steel with PCD tools. *Mater. Manuf. Processes* 18: 825–834. https://doi.org/10.1081/AMP-120024978.

52 Babitsky, V.I., Mitrofanov, A.V., and Silberschmidt, V.V. (2004). Ultrasonically assisted turning of aviation materials: simulations and experimental study. *Ultrasonics* https://doi.org/10.1016/j.ultras.2004.02.001.

53 Zhang, C., Ehmann, K., and Li, Y. (2015). Analysis of cutting forces in the ultrasonic elliptical vibration-assisted micro-groove turning process. *Int. J. Adv. Manuf. Technol.* https://doi.org/10.1007/s00170-014-6628-3.

54 Chang, S.S.F. and Bone, G.M. (2010). Burr height model for vibration assisted drilling of aluminum 6061-T6. *Precis. Eng.* 34: 369–375. https://doi.org/10.1016/j.precisioneng.2009.09.002.

55 Chang, S.S.F. and Bone, G.M. (2005). Burr size reduction in drilling by ultrasonic assistance, in: Robot. *Comput. Integr. Manuf.* https://doi.org/10.1016/j.rcim.2004.11.005.

56 Brehl, D.E., Dow, T.A., Garrard, K., and Sohn, A. (2006). Micro-structure fabrication using elliptical vibration-assisted machining (EVAM). In: *Proceedings of the 21st Annual ASPE Meeting*. ASPE.

57 Lin, J., Han, J., Lu, M. et al. (2017). Design, analysis and testing of a new piezoelectric tool actuator for elliptical vibration turning. *Smart Mater. Struct.* 26: 085008. https://doi.org/10.1088/1361-665X/aa71f0.

58 Kim, G.D. and Loh, B.G. (2010). Machining of micro-channels and pyramid patterns using elliptical vibration cutting. *Int. J. Adv. Manuf. Technol.* 49: 961–968. https://doi.org/10.1007/s00170-009-2451-7.

59 Tao, G., Ma, C., Bai, L. et al. (2017). Feed-direction ultrasonic vibration–assisted milling surface texture formation. *Mater. Manuf. Processes* 32: 193–198. https://doi.org/10.1080/10426914.2016.1198029.

2

Review of Vibration Systems

2.1 Introduction

A well-designed vibration system is quite important in vibration-assisted machining. A typical vibration system consists of vibration sources (actuators), a vibration transmission/amplification mechanism, and a control system. Generally, the actuators and transmission/amplification mechanisms were selected to be complementary. Given that certain demands such as required vibration frequency and amplitude range are proposed in the design phase, the optimal overall device structure to match those demands is determined first. Then, the key structural parameters are optimized using the finite element analysis (FEA) method, and the corresponding dynamic and static characteristics can be obtained. As a result, these key values in turn influence the choice of the vibration actuators. According to the system's operating frequencies, these proposed vibration devices can be divided to two groups: the nonresonant mode and the resonant mode. In a nonresonant vibration system, the vibration actuators usually vibrate below its first natural frequency. In order to increase the stability and reduce the dynamic error in the vibration stage, flexure hinge structures are widely used because of their superior dynamic response, low friction, and ease of control. For a resonant system, a sonotrode (also called a horn or concentrator) vibrates at its natural frequency, transferring and amplifying a given vibration from a vibration source, which is usually a magnetostrictive or piezoelectric transducer. This system can achieve a higher operating frequency and greater energy efficiency compared with a nonresonant system. However, its vibration trajectory cannot be controlled precisely owing to the nature of resonant vibrations and the phase lag between excitation and the mechanical response. Compared with resonant systems, nonresonant systems tend to achieve higher vibration accuracy, and it is easier to achieve closed loop control of the vibration trajectories under low-frequency conditions.

This chapter provides a general understanding of vibration systems, including actuator types and their selection, transmission/amplification mechanisms, and control system design. By analyzing design theories and principles, the advantages and disadvantages of various types of vibration devices are discussed. Moreover, future trends in vibration devices are also mentioned.

Vibration Assisted Machining: Theory, Modelling and Applications,
First Edition. Lu Zheng, Wanqun Chen, and Dehong Huo.

2.2 Actuators

As power output devices, actuators convert other types of energy into mechanical energy to drive the vibration stage. This contributes not only to the bandwidth of the vibration frequency and amplitude but also to the accuracy of the motion of the vibration stage. Currently, two types of actuators, namely, piezoelectric and magnetostrictive actuators, are mainly selected for vibration systems. This section provides a general understanding information about these actuators, including their strengths and weaknesses.

2.2.1 Piezoelectric Actuators

Piezoelectric actuators apply the unique piezoelectric properties of piezoelectric ceramics to convert high-voltage electrical energy into mechanical energy for high-frequency vibration. Compared with other types of actuators, they have the characteristics of low cost, simple structure, small size, fast response, high control precision, the lack of magnetic field and electrical fields, and no electromagnetic interference or electromagnetic noise, improving system stability and leading to more flexible designs of vibration devices. Various types of piezoelectric actuators, including stacked, thin plate, tubular, and bimorph types, have been developed for different applications [1, 2]. Considering factors such as displacement output, stiffness, and frequency response, stack-type piezoelectric actuators made of multiple pieces of piezoelectric ceramic plates mechanically connected in series and electrically connected in parallel are usually chosen in actual vibration systems.

2.2.2 Magnetostrictive Actuators

In the 1960/1970s, the technology of PZT was not yet fully understood. In addition, a magnetostrictive actuator was the best choice for vibration devices. It applies a ripple voltage to the electromagnetic coil, and its electromagnetic force is used to cause the moving core to vibrate. Although electromagnetic vibrators are a reliable source of vibration, the major drawback is low energy efficiency caused by high electrical eddy current losses. These electrical losses are transformed into heat and may damage the vibrator [3]. Therefore, with electromagnetic vibrators, the cooling issue always needs to be considered, leading to bulky size [4]. Currently, magnetostrictive actuators are usually applied in vibration systems that require low vibration frequency and large vibration amplitude.

2.3 Transmission Mechanisms

Two mechanisms using either flexure hinges or an ultrasonic horn are mainly chosen in the design of vibration system. The history of flexure hinge structures can be traced to the 1960s. With the development of the aerospace and aviation sectors, the resolution and size of the support in order to achieve small deflection ranges no longer met the requirements. After exploring various types of elastic support tests, engineers gradually developed flexure hinges that were characterized by high resolution, small volume, and no mechanical friction or gaps. Currently, many types of flexure hinges, including circular, elliptical, square fillet, and single-notch profiles, have been developed and are widely used to guide the displacement of vibration in the nonresonant vibration-assisted systems. In addition, they are often integrated into a double parallel or parallel four bar linkage so as to reduce the coupling motion because the nonresonant vibration stage is usually designed to work in two or more dimensions [5]. The ultrasonic horn is an important component of a resonant vibration system. It is used to transmit the mechanical energy converted from electrical energy into the workpiece by the transducer. It is a stage of the mechanical amplification of the power ultrasonic amplitude to improve the ultrasonic processing efficiency.

2.4 Drive and Control

Besides accuracy in the processing and assembly of mechanical components, the control strategy used also has a great influence on the motion accuracy of the vibration system. Generally, these control systems can be divided as open or closed loop systems. Because of the high working frequency involved, an open loop system is the first choice for a resonant vibration system. To build a proper open loop control system for a piezoelectric actuator, a mathematical model of the piezo-driven stage needs to be built first because of its hysteresis and nonlinear and creep properties. Many methods taking into account the intrinsic mechanism and dynamic properties of piezoelectric actuators have been developed in recent decades, including the Preisach, Maxwell, and Prandtl–Ishlinskii models [6–8]. The reference is calculated from the corresponding control signal according to the reference input value and it is sent to the piezoelectric actuator through a piezoelectric amplifier to generate a corresponding displacement. The features of open loop control system include a simple structure and ease of implementation; however, when the object or control device is disturbed or the characteristic parameters change during the working process, error cannot be compensated because this affects the accuracy of control. To overcome this drawback, closed loop control systems are used. Close loop control systems are mainly used in nonresonant vibration systems, and an industry standard controller is required. The traditional proportion integration differentiation (PID) control method algorithm has high control precision, but it is not suitable for uncertain time-varying systems. In contrast, fuzzy adaptive PID control can effectively identify the mathematical model of the controlled object, adjust the parameters and structure of the controller in real time according to the given performance indicators, and reduce the output error at this stage.

2.5 Vibration-Assisted Machining Systems

2.5.1 Resonant Vibration Systems

As a technology that has been successfully applied commercially, resonant vibration-assisted machining systems work at the natural frequency of the system and apply the excitation vibration principle to increase the amplitude of vibration. A typical design for an ultrasonic vibration-assisted machining system is the resonant rod type, which consists of three parts, namely, an ultrasonic transducer, an acoustic waveguide booster, and a

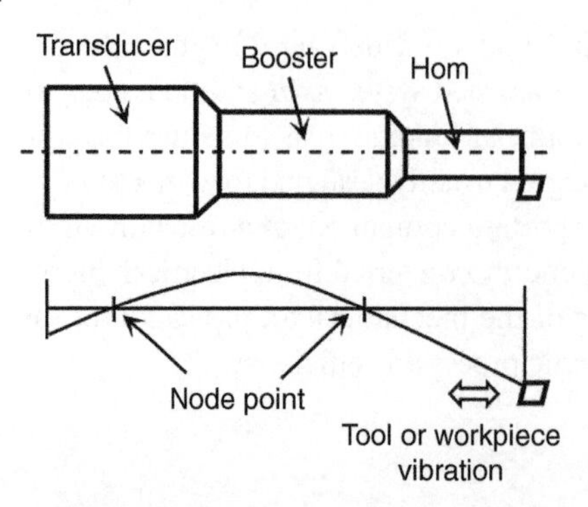

Figure 2.1 Typical design of a resonant vibrator.

horn (see Figure 2.1). In some research, the acoustic waveguide booster and horn are also called a sonotrode because the functions of the two components are quite similar [9]. The ultrasonic transducer is the source of vibration for the whole system and converts electrical energy into mechanical motion in longitudinal or compressive mode under self-excited vibration [10]. Two types of electromagnetic and piezoelectric transducers are widely used and were introduced in Sections 2.2.1 and 2.2.2. The high-frequency, low-amplitude reciprocating harmonic vibration is generated by the ultrasonic transducer and amplified by the sonotrode to the desired location of a tool or workpiece. The sonotrode works by resonating with the transducer, and there are strict requirements for its design and manufacturing. Poor design or fabrication will decrease the energy efficiency, reduce the cutting performance and vibration system durability, and may even cause serious damage to the transducer [11–14]. The cutting tool or workpiece is attached to the end of the horn to obtain the desired vibration. Moreover, the hold point of the whole system is usually set at the node point with zero displacement in order to maintain its stability and reduce energy loss. According to the direction of movement, a resonant vibration system can be divided into three groups: 1D, 2D, and 3D systems.

2.5.1.1 1D System

1D ultrasonic vibration-assisted machining systems are the most common type because of their simple structure and ease of implementation. They can be divided into resonant rod and resonant tool types. Many researchers have proposed their own rod type vibrators. Zhong et al. [15] improved the typical resonant rod-type system and applied it to the turning process, as shown in Figure 2.2. A tool holder with a notch structure is introduced into the design to hold the tool firmly in place and to reduce its moving in the other degrees of freedom. During the machining process, bending occurs at the notch point to prevent deformation in the rest of the tool holder. Otherwise, the tool holder in proximity to a parabolic shape will affect the machining performance. To obtain a more accurate measurement of cutting force during the vibration-assisted milling process, a special clamp system was designed by Shen et al. [16] by integrating the clamping system and a dynamometer, as shown in Figure 2.3. The results showed that the impact of ultrasonic vibrations on measurement results is reduced effectively. Similarly, Liu et al. [17–19] studied ductile mode cutting with tungsten carbide, as shown in Figure 2.4. The new clamping system fixes the vibrator using four bolts, which simplifies the installation procedure of the vibrator and improves its accuracy.

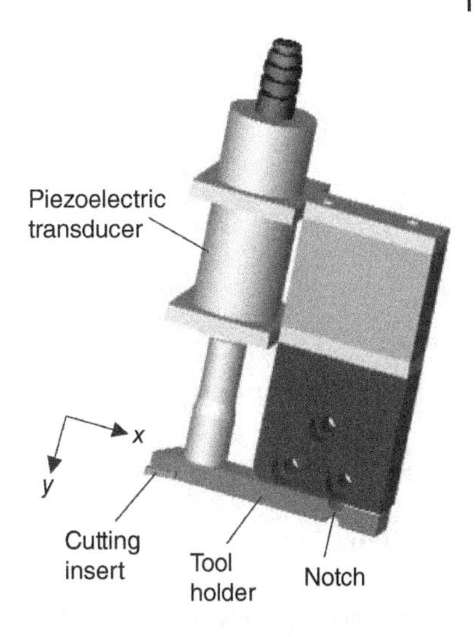

Figure 2.2 Vibrator proposed by Zhong et al. Source: Zhong et al. [15].

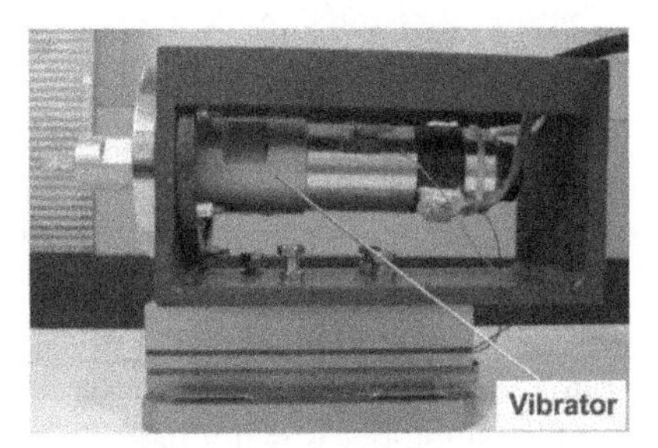

Figure 2.3 Vibrator proposed by Shen et al. Source: Shen et al. [16]. © 2012, Springer Nature.

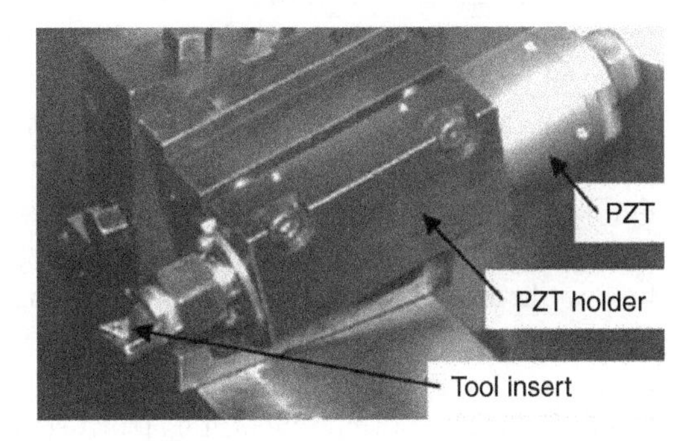

Figure 2.4 Vibrator proposed by Liu et al. Sources: Liu et al. [17–19].

The resonant rod-type 1D resonant vibration system has a simple structure and high reliability; however, the resonance frequency of the system can be easily influenced when a workpiece or a large mass is attached to the horn. Meanwhile, the issue of the installation of oversized part is also difficult to solve. Therefore, another type of resonant vibration system named resonant tool was developed by integrating the resources of vibration into the tool holder. A typical design was proposed by Ostasevicius et al. [20]. The milling cutter assembly is driven by piezoceramic rings that are fixed into a standard Weldon tool holder and generate resonant tool movement in the vertical direction. Similarly, Alam et al. [21]

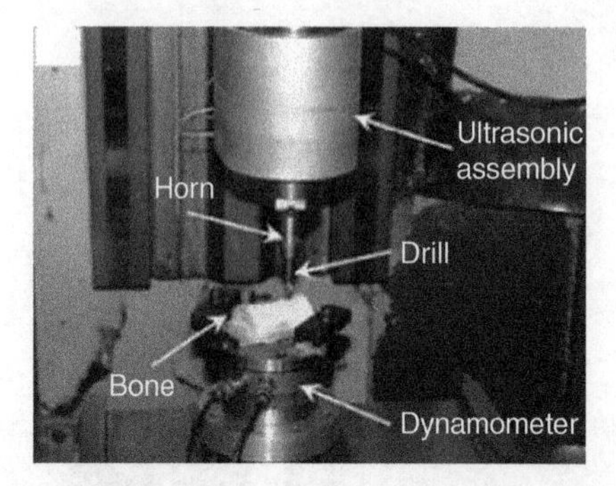

Figure 2.5 Vibrator proposed by Alam et al. Source: Alam et al. [21]. © 2011, Elsevier.

improved the tool cutting assembly design and obtained a sevenfold increase in vibration amplitude by using a stepped shape of horn structure (Figure 2.5).

As discussed in the previous sections, the vibration parameters for an ultrasonic vibration system largely depend on the dimensions and cross-sectional shape of the designed vibration transmission mechanism consisting of the booster and horn. However, the traditional approach is based on the application of differential equations where the equilibrium of an infinitesimal element is taken into consideration under the influence of elastic and inertia forces. This is time-consuming and inaccurate. To overcome these drawbacks, FEA is introduced at the design stage of the ultrasonic vibration system, and its use can increase the accuracy of the vibration system, such as in natural frequency and the dimensions of the mechanism, which speeds up the development of vibration devices. Kuo [22] proposed a milling cutter assembly design where the process of harmonic piezoelectric vibrations was simulated by an FEA dynamic simulation, which optimized the key dimensions, reduced the influence of stress concentration on the system, and increased its system efficiency. However, the simulation did not consider a situation where a tool is attached to the horn, and this leads to a deviation in the system's natural frequency and vibration amplitude between the simulation results and operational results. Roy et al. [23] developed a circular hollow ultrasonic horn for milling cutter assembly and optimized its outline and cross-sectional shape by using FEA. Compared with conventional ultrasonic horn designs, such as those with stepped, conical, and exponential shapes, the circular hollow ultrasonic horn achieved a higher magnification factor and lower axial, radial, and shear stress, hence improving the system performance and reducing the influence of stress concentration.

A different type of vibration drilling tool assembly design was proposed by Babitsky et al. (Figure 2.6). In order to accomplish vibration-assisted drilling, one side of the assembly was clamped in the three-jaw chunk of the lathe through the intermediate bush and energized by means of a slip ring assembly fitted to the hollow shaft of the lathe at the end remote from the chuck [24]. A further 1D ultrasonic vibration system was developed by Hsu et al. [25], and its working principle is quite similar to the ultrasonic bath. As shown in Figure 2.7, three commercial Langevin ultrasonic transducers were fixed underneath the vibration stage and were controlled by the same type of signals, generating vibrations at the same frequency and phase. As a result, resonance vibration can be obtained in the vibration stage.

Figure 2.6 Vibrator proposed by Babitsky et al. Source: Babitsky et al. [24]. © 2007, Elsevier.

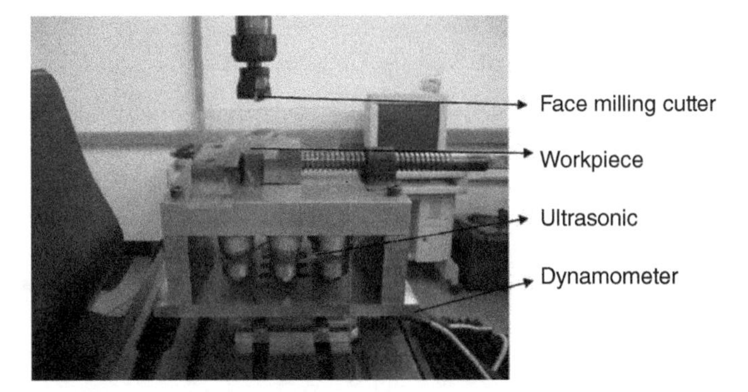

Figure 2.7 Vibrator proposed by Hsu et al. Source: Hsu et al. [25]. © 2007, Springer Nature.

2.5.1.2 2D and 3D Systems

Ultrasonic elliptical vibration-assisted machining is also named 2D ultrasonic vibration-assisted machining and has received widespread attention since it was first proposed in 1993. Compared to 1D systems, 2D systems can obtain better cutting performance and also require a higher standard of vibration devices. Because of its simple structure and ease of implementation, the integrated resonant rod device is the most popular in proposed 2D vibration devices. There are two main designs: patch and sandwich types. For the patch type of integrated resonant rod, two sets of piezoelectric plates are attached to the outer wall of the resonant rod to achieve the same or different modes of resonance. In the sandwich type, two modes of resonance moment can be obtained by adding another set of piezoelectric rings to the 1D resonant rod. Moriwaki et al. [26] developed a 2D patch-type ultrasonic vibration-assisted turning system (Figure 2.8) by attaching two pairs of piezoelectric actuators symmetrically in the center of the four sides of a stepped horn. When two sinusoidal signals with different phases and the same frequency are applied to the piezoelectric actuators, a bending resonance state can be obtained in the vibrator in two mutually perpendicular directions simultaneously and the cutting tool attached at the end of the vibrator vibrates in elliptical mode. The coupling effect cannot be avoided, as the piezoelectric plates are placed in parallel. Shamoto et al. [27, 28] optimized the dimensions and shape of the vibration rod and developed a control system to achieve a more accurate tool trajectory. Experiments were conducted on hardened stainless steel and the machining accuracy and

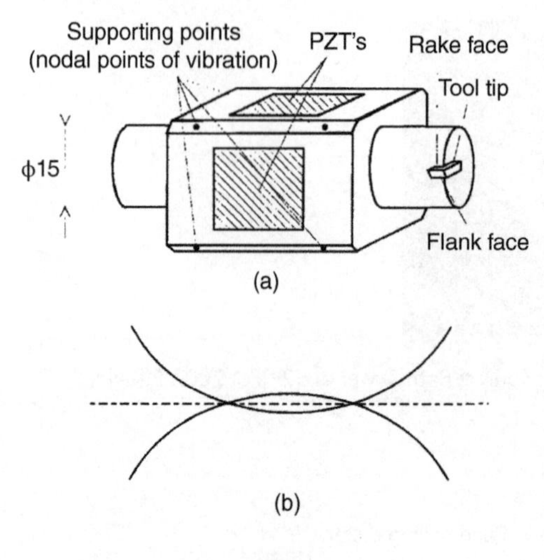

Figure 2.8 Vibrator proposed by Moriwaki and Shamoto. (a) Ultrasonic vibrator and (b) first resonant mode of bending. Source: Moriwaki and Shamoto [26].

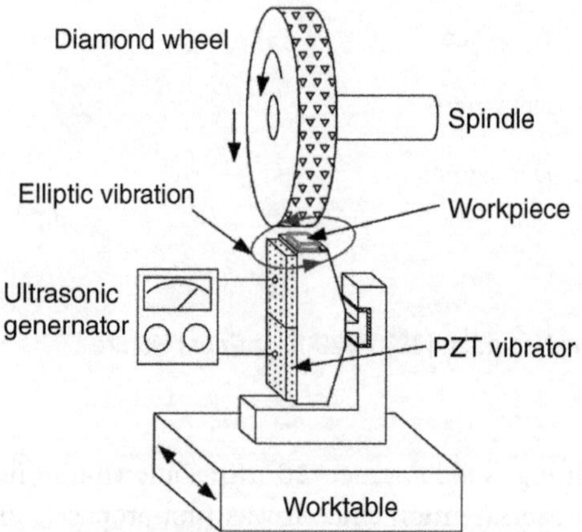

Figure 2.9 Vibrator proposed by Liang et al. Sources: Liang et al. and Peng et al. [29–31].

machined surface quality improved. A combination of bending and longitudinal vibration modes was also achieved in a different 2D patch-type ultrasonic vibrator designed by Liang et al. [29–31], as shown in Figure 2.9. The piezoelectric plates are bonded at the same side of the metal elastic body, and the workpiece is fixed to the top of the vibrator. However, the vibration amplitudes involved are quite small, only up to 0.4 μm, because of the mass issues with the vibrator, and an effect of workpiece mass on the vibration's amplitude and frequency is unavoidable.

A typical sandwich-type elliptical ultrasonic vibrator is shown in Figure 2.10 [32, 33]. The two groups of piezoelectric rings are sandwiched together in the drilling cutter assembly, and each group works at a different resonant mode to generate an elliptical tool tip trajectory. It should be noted that the installation point of this type of vibration device should be set at the coincidence point of the two resonance modes. Meanwhile, vibrators for mounting workpiece or nonrotating tools such as for turning and polishing have also been proposed using the same design principle [34–39]. To achieve better performance, Börner et al. [40] developed a cross-shaped converter for a 2D ultrasonic vibration-assisted vibrator (Figure 2.11) and applied it to the milling process. As the high-frequency vibration

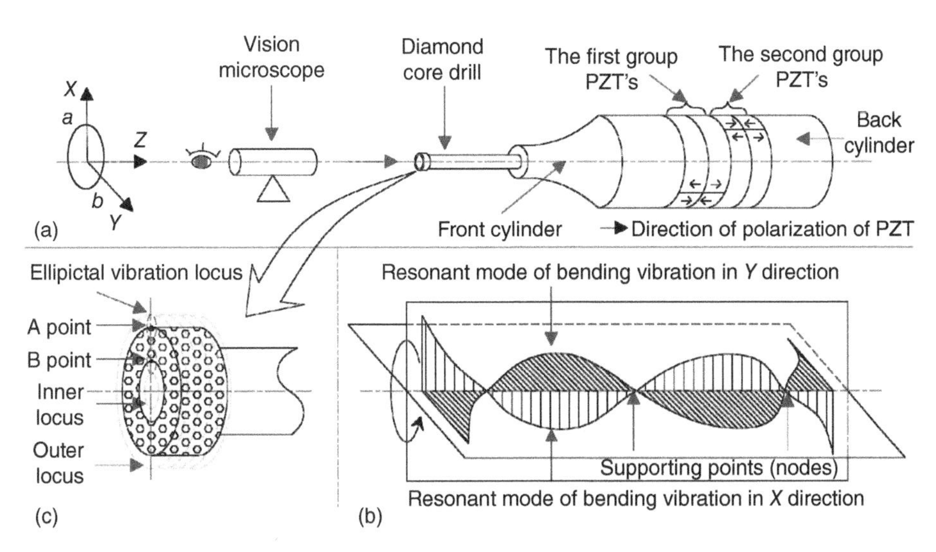

Figure 2.10 Vibrator proposed by Liu et al. Source: Liu et al. [32].

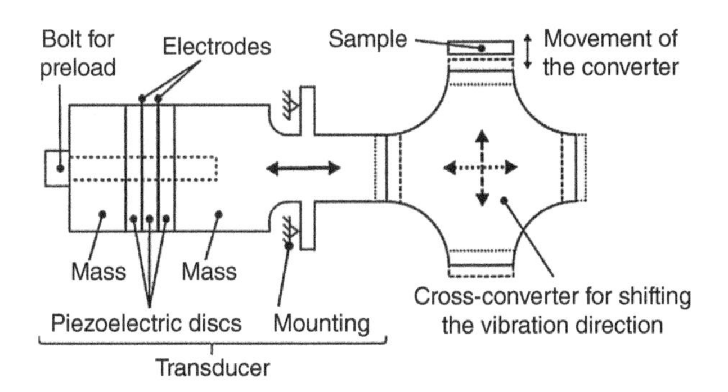

Figure 2.11 Vibrator proposed by Börner et al. Source: Börner et al. [40].

is transmitted to the cross-converter, the extension in horizontal direction will lead to a compression in the vertical direction. However, only small specimens can be used because those with large mass may influence the resonance frequency of the vibrator. Tan et al. [41] built a symmetrically structured ultrasonic elliptical vibration-assisted device (Figure 2.12) using four pairs of piezoelectric rings. The node point of the device is naturally set at the center of the flange, which is easy to locate. The device is fixed by two pairs of grip holes displaced on both sides of the flange to reduce the energy loss and improve the cooling performance. Compared with the conventional design, the symmetrical structure can completely balance the internal force, which dramatically reduces error in the vibrator's motion. By changing the piezoelectric actuator to a magnetostrictive actuator, Suzuki et al. [42–44] developed an elliptical vibrating polisher (Figure 2.13) and successfully applied it to process micro-aspheric lenses and tungsten carbide die/mold. The vibrator is based on a giant magnetostrictive material and with coils wound around it. It has four legs and each leg can be independently controlled for expansion or contraction. The elliptical tool trajectory can be obtained by appropriately setting the phase difference of the two pairs of opposing coils.

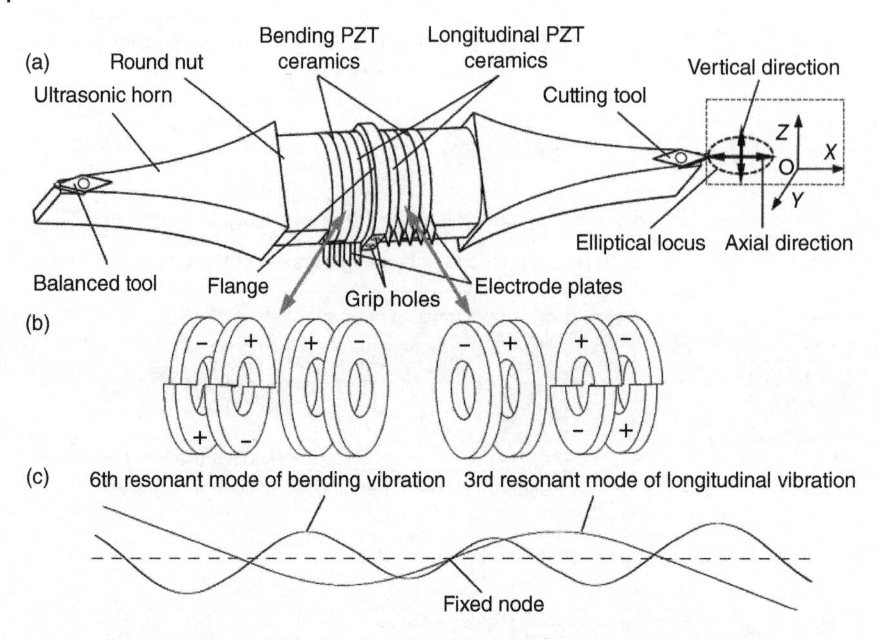

Figure 2.12 Vibrator proposed by Tan et al. Source: Tan et al. [41].

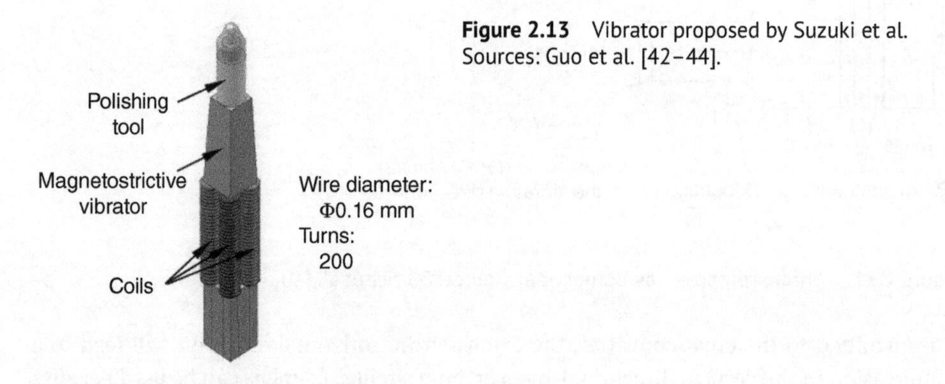

Figure 2.13 Vibrator proposed by Suzuki et al. Sources: Guo et al. [42–44].

To obtain the highest vibration amplitude, the vibrator is designed to work at a frequency of 9.2 kHz, which limits its processing performance.

Different from integrated resonant mode vibration devices, a separate type of 2D resonant vibrators uses two independent Langevin vibrators placed in a V or L shape to obtain a two-dimensional vibration of the tool or workpiece [45–47]. Figure 2.14 shows a typical separate 2D V-shaped vibrator proposed by Guo et al. [48]. The two Langevin vibrators are set at an angle of 60° to generate a unique tool tip trajectory. The head block is a flexure structure applied at the end of each vibrator to guide motion and reduce movement error. Each individual vibrator has an added end mass to preload the piezoelectric rings and adjust the natural frequency of the vibrators. A similar design was also applied in a vibration-assisted polishing process [49]. Yan et al. [50] developed a 2D "L"-shaped resonant vibrator for grinding, as shown in Figure 2.15. Two independent 1D resonant vibrators are placed perpendicularly on the sides of the vibrating stage. However, this type of 2D

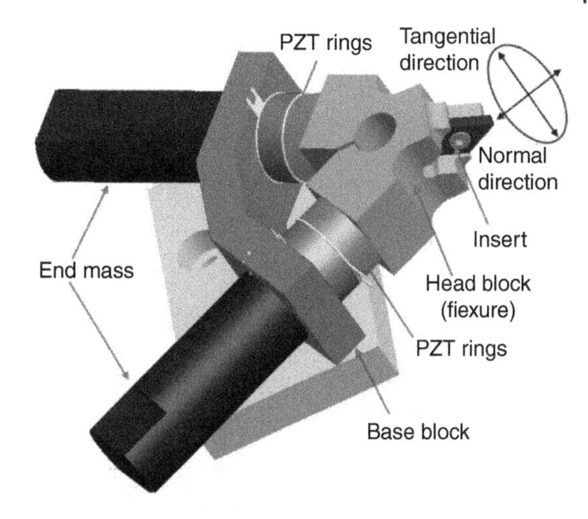

Figure 2.14 Vibrator proposed by Guo and Ehmann. Source: Guo and Ehmann [48].

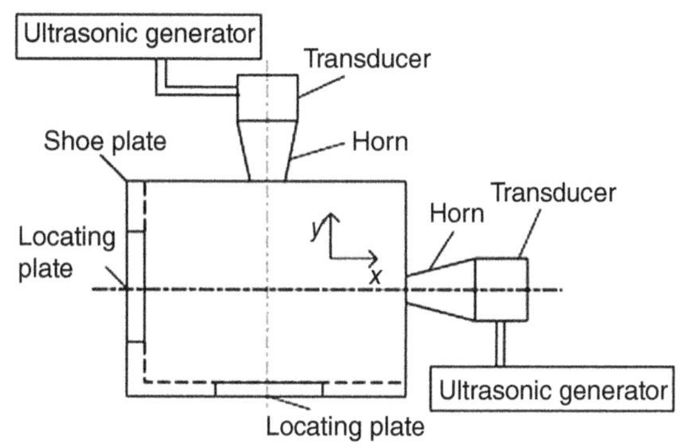

Figure 2.15 Vibrator proposed Yanyan et al. Source: Yanyan et al. [50].

resonant vibrator is almost impossible to integrate into a rotating tool such as in a milling cutter assembly, which limits its application.

In order to obtain complex geometrical shapes, the milling process requires a feed vector in arbitrary directions with both vertical and horizontal components of feed vector necessary for 3D end milling. Hence, there is a need for three-dimensional vibration assistance. Figure 2.16 shows a three degree of freedom resonant vibration tool, which can generate longitudinal and two bending resonance mode vibrations by adding three sets of piezoelectric actuators to a resonance rod [51–53]. The difficulty associated with this design is to accurately locate the node point of the vibration rod and to achieve three modes of resonance frequencies, which are as close as possible to obtain sufficient vibration amplitude. Moreover, the cross talk between the three resonance modes is much more prominent than that in a 2D resonant system. In order to reduce the motion coupling effect, stepped and the tapered portions are added to the resonant rod, and the overall shape and dimensions are optimized so as to obtaining optimal performance.

2.5.2 Nonresonant Vibration System

Generally, resonant vibration systems are capable of achieving extremely high operating frequencies, and most of them can even reach ultrasonic vibration frequency levels ($\geq$20 kHz).

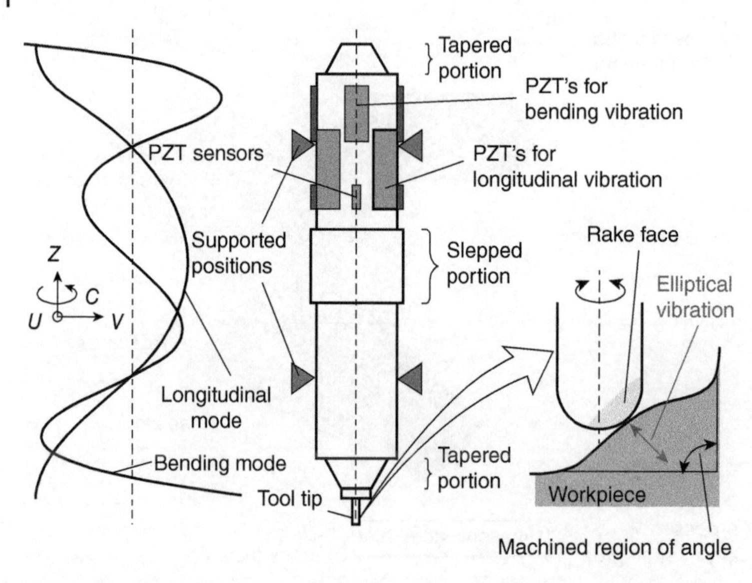

Figure 2.16 Layout of the 3D vibrator. Sources: Suzuki et al. and Shamoto et al. [51–53].

However, their limitations of fixed working frequency and vibration motion parameters, heat dissipation, open loop control, and poor dynamic accuracy are also quite obvious. In addition, the performance of the vibrator heavily relies on the dynamic characteristic of the vibration horn, which increases the difficulty of vibrator design. To overcome these shortcomings in the resonant vibrator, much more attention has been paid to nonresonant vibration systems. Nonresonant systems apply forced vibration rather than excitation vibration as the design principle and produce variable vibration frequencies. However, it is hard to achieve a high working frequency, which is always less than the natural frequency, because of the issue of structural stiffness. Many of these designs are inspired by high-precision micro/nano-positioning stages [54, 55], which are discussed below.

The working principle of a nonresonant vibration system can be explained by the schematic diagram in Figure 2.17. The whole system is driven directly by the preloaded piezoelectric actuator. In order to accurately transmit motion and reduce parasitic movement, flexure mechanisms (flexure hinges), which can be simplified into a set of spring–damper mechanisms, as shown in Figure 2.17, are always chosen as the linkage between the actuators and end executor. A displacement amplifying mechanism can be integrated into the flexure hinges if the amplitude is required larger than the displacement generated by the piezoelectric actuator. Moreover, decoupling issues also need to be

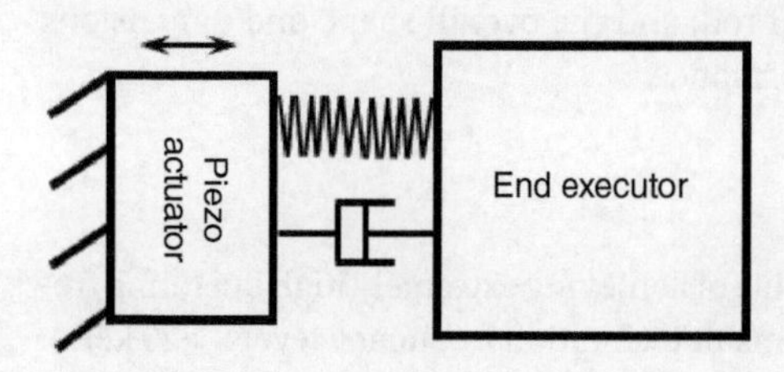

Figure 2.17 Typical working diagram of a nonresonant vibrator.

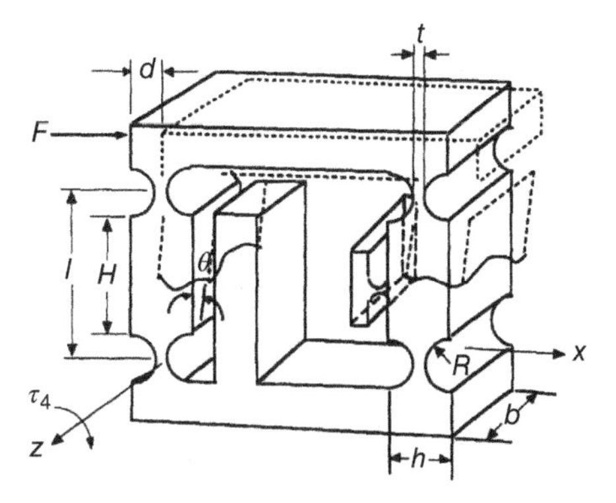

Figure 2.18 Vibrator proposed by Greco et al. Source: Hong and Ehmann [57].

considered in the design phase of the flexure mechanism when multidimensional motions are required.

Compared with resonant vibration systems, higher motion accuracy can usually be achieved with a nonresonant vibration system because of its inherent merits. This makes it more suitable for the manufacturing of microstructured surfaces. Therefore, 1D nonresonant vibration systems are quite rare because of the complex tool trajectories required in producing unique surface microstructures. A typical design of a 1D nonresonant vibration system uses a combination of a parallel four-bar flexure hinge structure and a piezoelectric actuator [56]. Figure 2.18 shows a 1D nonresonant vibration system design proposed by Long et al. [57, 58]. A single piezoelectric actuator is positioned at the parallel four-bar flexure hinge structure, which also includes a vibration displacement amplification function. The structural layout and closed loop control system ensure high motion precision. A different design was proposed by Suzuki et al. [59] with a complex mechanical structure. To ensure cutting accuracy, this vibrator aims to achieve high axial mechanical stiffness so as to reduce elastic deflection during the machining process. A cylindrical roller bearing is set between the cutting tool and piezoelectric actuators to guide the vibration and support the bending force acting on the cutter insert, and the twisting force in the machining process is supported by the pin. Moreover, bending stress is further reduced because of the flexible tip. Consequently, the shear stress that could damage them cannot be transmitted to the piezoelectric actuators. Because the output voltage for the control of the vibrator may reach up to 1000 V, an air-cooling system is also integrated into the vibrator to prevent overheating.

2.5.2.1 2D System

Compared with 1D systems, the application of 2D systems is more flexible. However, the coupling effect between the two vibration directions has a great impact on its accuracy. Two configurations exist among the proposed designs, a vibration tool mainly for turning and a vibration stage mainly for milling. A flexure hinge structure is often used to guide motion and to reduce motion error and the coupling effect, although some other designs have also been reported. Brehl et al. [60–62] developed two types of nonresonant elliptical vibration-assisted machining devices at two different working frequencies. One of these vibrators (Figure 2.19a) works at a high vibration frequency of up to 4.5 kHz but a low vibration amplitude of less than 2 μm and requires a cooling chamber to prevent the vibrator

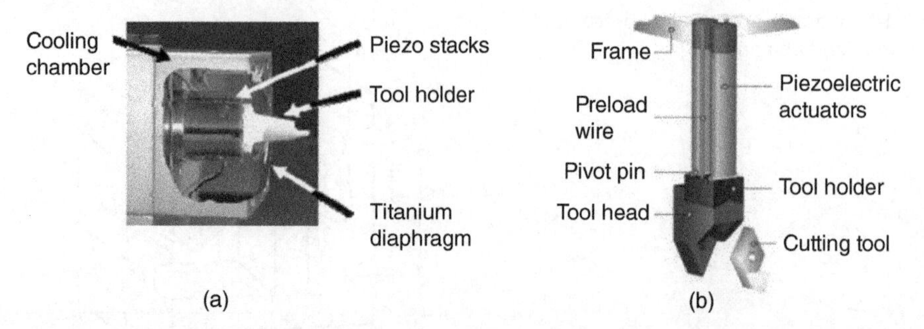

Figure 2.19 Vibrator proposed by Brehl et al. Sources: Brehl et al. [60–62]. © 2006, ASPE.

from overheating. The other design (Figure 2.19b) operates at a low vibration frequency of 400 Hz but high vibration amplitude up to 22 μm and does not need an extra cooling system. The design principles of the two vibrators are almost the same. Two piezoelectric actuators are placed in parallel and generate two sinusoidal vibrations with a specific phase difference. Then, two sets of parallel high-frequency vibration motions are converted by the "T" shape tool holder into the desired elliptical cutting tool trajectory. To simplify the structure and improve the working frequency band of the vibrators, the piezoelectric actuators are used to achieve cutting tool support and positioning in both vibrator designs. However, the drawback of this design is that the accuracy of the vibrators is hard to guarantee under the conditions of open loop control. Therefore, it is not suitable for the precision machining processes because of the nonlinear hysteresis and creep behaving of the piezoelectric actuators, which is unavoidable.

Another two designs of 2D nonresonant vibration tool assembly (Figure 2.20) were proposed by Kim et al. [63–65]. Two stacked piezoelectric actuators are fixed/preloaded by a through bolt on the tool holder and placed in parallel and perpendicular to each other, respectively. A similar design proposed by Li et al. [66] uses four pairs of piezoelectric plates symmetrically placed at both sides of the vibration platform. However, the major drawback of these designs is the motion coupling between the two motion directions, which

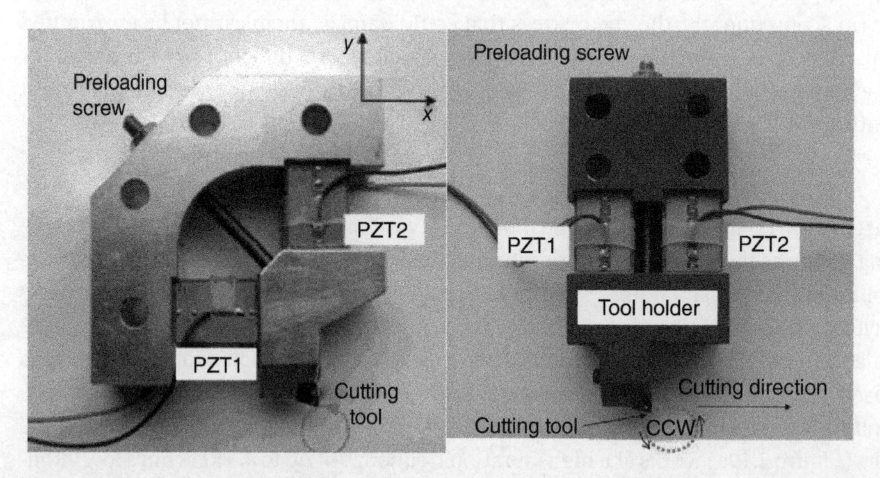

Figure 2.20 Vibrator proposed by Kim et al. Sources: Kim et al. [63–65]. © 2010, Springer Nature.

causes tangential shear stress inside the piezoelectric actuator and damages the vibrator. To overcome these drawbacks, many optimized design proposals have been proposed. Loh et al. [67] improved the current vibrator design, and a better performance was obtained in their experimental results. Compared with the original design, the structure of the tool holder is optimized, and a hemispherical structure is added at the junction of the piezoelectric actuator and tool holder, which changes the connection mode between them from surface contact to point contact. As a result, the motion coupling between the two piezoelectric actuators can be effectively avoided and potentially hazardous shear deformation of the piezoelectric actuators reduced. Another design adds cross-shaped voids to the tool vibrator, and the results show that the cross-interference of the vibration of each axis is effectively suppressed [68]. Chern et al. [69] put forward a different vibration design employing a combination of piezoelectric actuators and linear guideways (see Figure 2.21). However, the coupling effect cannot be reduced effectively because the end spring of the vibrating stage has no torsional stiffness.

To obtain better cutting performance and a more accurate tool tip trajectory, various nonresonant vibration stages with flexure structures have been proposed. One of these designs uses independent driving, where each group of piezoelectric actuator drives one direction of the vibrator and movement in each direction is independent of the others in order to minimize mutual coupling. A typical 2D nonresonant vibration stage based on the flexure structure was developed by Jin et al. [70]. The movable vibration table is connected to the piezoelectric actuators through flexure hinge structures in both directions. As the two piezoelectric actuators work together, the motion of the flexure hinges will be affected by coupling stiffness, leading to an uneven displacement between the two directions, which makes it hard to control them precisely. Ding et al. [71, 72] built a nonresonant elliptical vibration stage using a parallel kinematic double-flexure hinge structure. The stage is driven by two independent piezoelectric actuators, which are placed perpendicular to each other. The parallel kinematic double-flexure hinge structure has advantages such as compactness, high stiffness levels, and no friction. In addition, it also reduces the coupling effect of the vibration stage. Also, a single flexure parallel four-bar elliptical vibration stage has been developed using four flexure hinges supporting the movable stage (see Figure 2.22) [73]. However, a large stress concentration and parasitic movements cannot be eliminated because of structural defects. A planar integrated vibration stage with a symmetrical double flexure parallel four-bar structure was proposed by Zhang et al. [74]. It features ax compact structure, zero clearance, and no mechanical friction, which leads to better

Figure 2.21 Vibrator proposed by Chern et al. Source: Chern et al. [69]. © 2006, Elsevier.

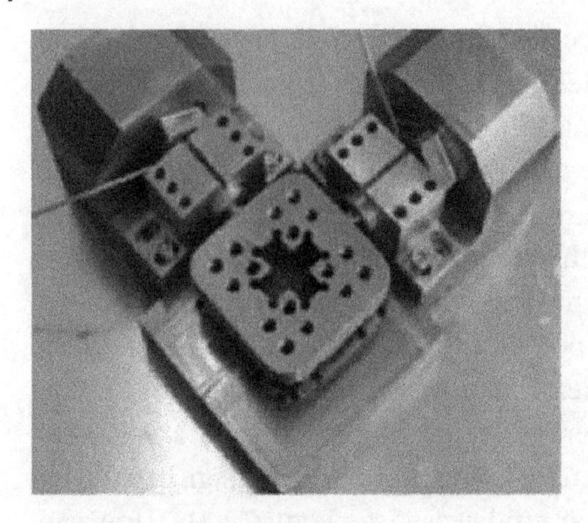

Figure 2.22 Vibrator proposed by Li et al. Source: [73]. Public Domain 2018, Springer Nature.

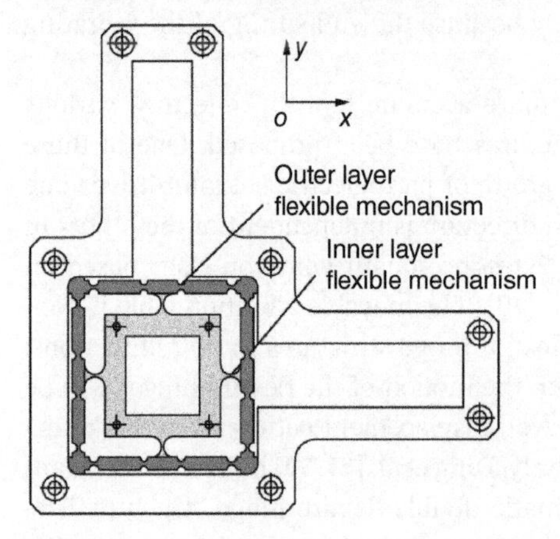

Figure 2.23 Vibrator proposed by Chen et al. Sources: Zheng et al. and Chen et al. [75–77].

guidance properties. Compared with the single-parallel four-bar structure vibration stage, it not only eliminates the cross-coupling displacement but also weakens the external interference to a certain extent. By optimizing the existing designs, Chen et al. [75–77] reported a 2D highly dynamic horizontal nonresonant vibration-assisted milling system. As shown in Figure 2.23, a structure of double-parallel four-bar linkages with double-layer flexible hinges is applied to the vibration stage design, which not only reduces the coupling effect but also improves the efficiency of the vibration transmission. The stiffness and displacement of the vibrating stage are assured by adjusting the dimensions of the different flexible hinges. Another compact 2D non-resonant vibration stage design was proposed in [78, 79]. Two pairs of piezoelectric actuators are placed symmetrical in flexible hinges around the oval mechanical structure. The circuit board is integrated into the vibration stage, which reduces the device's volume.

Another design applies two or more piezoelectric actuators in the same or multiple directions to produce a compound elliptical motion. A more complex tool trajectory can be obtained compared with an independently driven vibrating stage. In order to reduce the influence of other factors on the vibration amplitude and frequency, it is usually integrated with the machine tool [80]. A flexible hinge structure-based 2D vibration turning tool was also proposed as shown in Figure 2.24 [81]. The design principle of the vibrator uses a set of vertically connected biaxial flexible hinges to support and guide

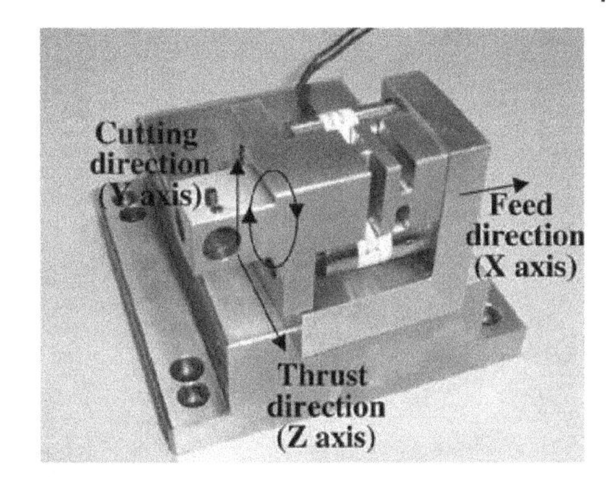

Figure 2.24 Vibrator proposed by Kim et al. Source: Kim et al. [81]. © 2009, Springer Nature.

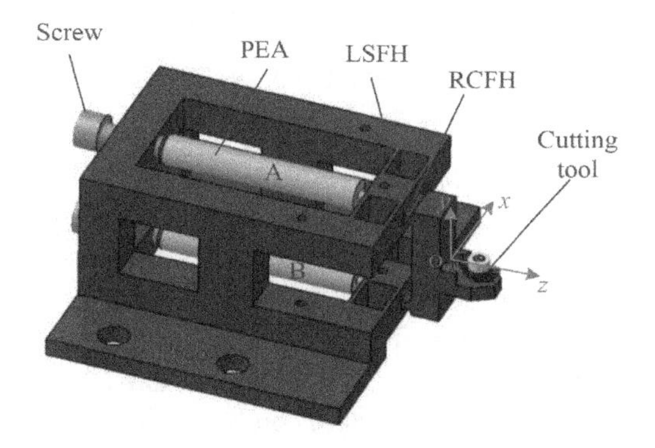

Figure 2.25 Vibrator proposed by Lin et al. Source: Lin et al. [82].

the tool holder. When the tool holder is driven by two piezoelectric actuators, the elastic deformation of the flexible hinges in different directions can be employed to synthesize the motion of elliptical tool tip. Lin et al. [82] reported a piezoelectric tool actuator (PETA) for elliptical vibration turning based on a hybrid flexure hinge connection (see Figure 2.25). It consists of two parallel four-bar linkage mechanisms and two right circular flexure hinges and effectively improves the vibration accuracy and reduces the coupling effect between the two motion axes. The step responses, motion strokes, vibration resolutions, parasitic motions, and natural frequencies of the PETA along the two input directions were analyzed, and the results show that the vibrator is capable of precision machining. A 2D low-frequency vibration-assisted polishing tool assembly is proposed by Chee et al. [83]. It consists of four mechanical amplitude-magnified actuators screwed together on a center piece, which generates a planetary elliptical tool tip trajectory. Although the nonintegral structure reduces the design and manufacturing difficulty of the vibrator, higher installation accuracy is also required because of the fact that the assembly accuracy has a great influence on the vibrator motion trajectory. To solve this problem and achieve higher tool motion accuracy, another 2D low-frequency vibrator is shown in Figure 2.26 [84]. Three mechanical amplitude-magnified actuators are arranged in a triangle around the centerpiece, and a highly repeatable and stable polishing trajectory can be obtained. In addition, the 2D low-frequency vibrator is attached to a low-contact force loader, which further reduces the polishing force and prevents damage to the surface layer of the workpiece. Compared with the previous system, this vibration system has advantages such

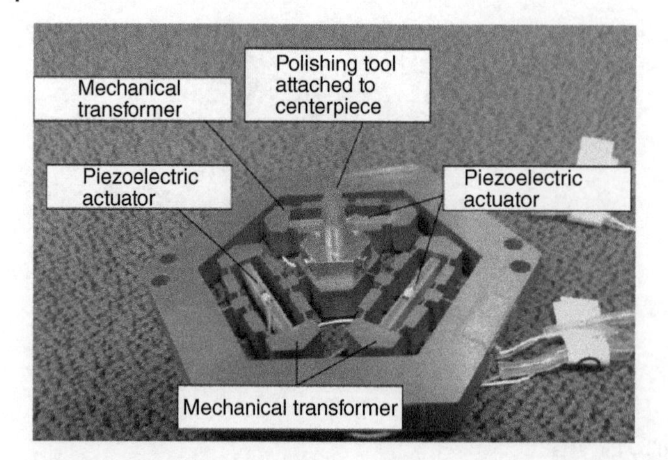

Figure 2.26 Vibrator proposed by Chee et al. Source: Chee et al. [84]. © 2013, Fuji Technology Press, Ltd.

as higher vibration accuracy and lower grinding force and energy consumption, and the resolution of the vibration is also improved to 0.1 μm.

2.5.2.2 3D Systems

As an important part of nonresonant systems, the 3D vibration system can generate a more complex tool trajectory that is suitable for special applications such as the production of optical freeform surfaces on hard and brittle materials [85]. Similar to the design of the 2D nonresonant vibrator, the 3D nonresonant vibrator can also be divided into compound motion and independent drive types. For the latter 3D nonresonant vibrator, the coupling can seriously influence the accuracy of motion because of the motion of the vibration device, which is composed of individual movements in three axes. In addition, the totally decoupled motions always require a large number of linkages with special structures needed for motion isolation, causing relatively low response rates [86]. Wang et al. and Liu et al. proposed two similar 3D nonresonant vibration devices [87, 88]. However, their complex flexible structures result in low bandwidth and response speeds. To overcome these drawbacks, a different type of independent drive 3D nonresonant vibration device has been developed, which is shown in Figure 2.27 [89]. In this design, low coupling and high motion

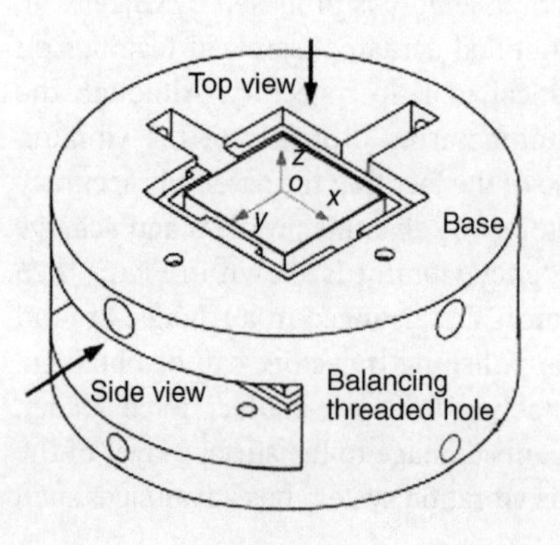

Figure 2.27 Layout of a 3D nonresonant vibrator. Source: Zhu et al. [89].

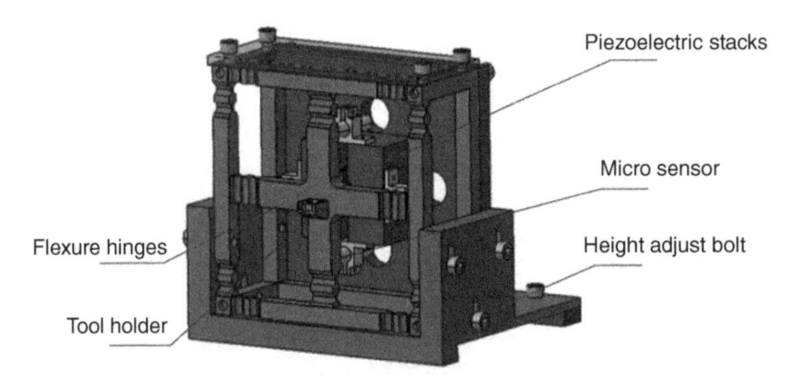

Figure 2.28 Compound motion-type 3D vibrator. Sources: Lu and Lin et al. [90, 91].

accuracy and response rates can be obtained by employing a compact and rotationally symmetrical structure and special flexure hinges with multiple degree of freedom (DoFs). Figure 2.28 shows a representative design of the compound motion-type 3D nonresonant vibration device [90, 91]. The vibration device is driven by four piezoelectric actuators, and the different 3D motion planes and trajectory can be obtained by changing the acting locations of the piezoelectric stacks or parameters of the signals. By installing displacement sensors, this vibration device achieves closed loop control and better motion accuracy.

2.6 Future Perspectives

With a deeper understanding of the cutting mechanisms in vibration-assisted machining technology, its potential for processing a hard and brittle material is gradually being explored. However, this also leads to stricter requirements for stability and real-time frequency tracking technology in vibration-assisted machining systems because of the high loads involved in the machining process. Therefore, an inevitable trend has been to improve resonant vibration systems to adapt to large loads and frequency drift. At present, ultrasonic resonance devices are basically set at the cutter side because of the unpredictable mass and size of workpieces, which further affects the amplitude and frequency of the resonant vibration. Nevertheless, the resonant state of the ultrasonic system is greatly influenced by the factors such as tool materials and clamping mode. Meanwhile, besides high-frequency ultrasonic vibration, high cutting loads are also applied directly to the cutter, which leads to high standard requirement of cutting tool [92]. Accordingly, one of the most important trends in resonant vibration-assisted machining systems is improvement in the system and the development of special tools. By taking into account the characteristics of ultrasonic vibration cutting and the integration of cutting tool and ultrasonic vibration system from the perspective of integrated collaborative design, tools can be adapted to the vibration system while meeting the processing requirements, which further ensures the stability of the resonant vibration system. In addition, the issue of overheating in ultrasonic vibration systems operating for long periods is also a problem, which needs an urgent solution. Designing a dedicated vibration system with an internal cooling function will help to solve this problem.

As reviewed in the previous section, 2D vibration-assisted machining systems are the most common type of nonresonant vibration systems, and these systems are always built using piezoelectric actuators to drive the flexible hinges directly. In addition, most flexible hinge structures are designed to be symmetrical and the current effort in performance improvement of nonresonant vibration-assisted machining systems is mainly focused on the optimization of shape, which limits improvements in performance. For serial flexible hinge structures in nonresonant vibration system, the main drawbacks are the accumulation of vibration error, low working frequency bandwidth, and structural redundancy. For parallel flexible hinge structures in nonresonant vibration systems, the coupling effect between different axes is unavoidable, which seriously affects the motion accuracy of the system and can induce shear stress inside the piezoelectric actuators and damage them. Therefore, structural optimization of nonresonant vibration systems so as to improve vibration accuracy and reduce coupling effect is a major development trend.

Currently, vibration-assisted machine tools for difficult-to-machine materials have not yet reached maturity. The reliability of the machine tool system and stability in long-term operation needs to be evaluated and strengthened. Many sectors such as the aerospace and medical areas have high demand for stable ultrasonic-assisted machine tools. Hence, integrating current research results for ultrasonic vibration-assisted technology and the development of special ultrasonic vibration auxiliary equipment will soon become important research priorities.

2.7 Concluding Remarks

From the ongoing in-depth research, most of the merits of vibration-assisted machining are being discovered, which extend its application in precision machining processes with hard and brittle materials. However, well-designed vibration devices are required to ensure their performance. In general, design should not only consider the operating frequency and amplitude but also the accuracy, ease of control, and application environment. This chapter has critically reviewed research and developments in vibration devices. It provides a discussion of both the advantages and disadvantages of existing vibration devices and gives insights into vibration device design in the future. The following are the key observations and suggestions for future work on vibration devices:

(1) Vibration-assisted machining has been in development for about 60 years since it was first proposed. Advantages such as extended tool life and better surface finishes have been discovered and have played an important role in improving the machining accuracy of hard and brittle materials and reducing costs. The sources of vibration have undergone on evolution from electrohydraulic actuators to electromagnetic actuators to piezoelectric actuators, and each evolution improves the motion accuracy of the vibration system.
(2) Resonant vibration devices usually vibrate at their natural frequency, and the vibration frequency and amplitude are fixed at certain values. Given the high demand for the precision machining of hard and brittle materials, high vibration frequency, low amplitude, and the undertaking of large loads will be the development trends for resonant vibration devices in the near future.

(3) The applications of nonresonant vibration systems are more flexible compared with resonant systems because of the variable vibration amplitudes and frequencies. Besides pursuing a wider range of working frequencies and amplitudes, a reduction in the coupling effect is also a significant mainstream research aim with this type of vibration device.

References

1 Bansevicius, R. and Tolocka, R.T. (2002). Piezoelectric actuators. In: *Mechatronics Handbook* (ed. R.H. Bishop), 20–51. Washington, D.C.: CRC PRESS Boca Raton London New York.

2 Crawley, E.F. and De Luis, J. (1987). Use of piezoelectric actuators as elements of intelligent structures. *AIAA J.* https://doi.org/10.2514/3.9792.

3 Geoffrey, B. and Winston, A.K. (1989). *Fundamentals of Machining and Machine Tools.* Marcel Dekker, Inc. https://doi.org/10.1007/978-1-84800-213-5.

4 Thoe, T.B. (1998). Review on ultrasonic machining. *Int. J. Mach. Tools Manuf.* 38: 239–255.

5 Wang, G., Zhou, X., Ma, P. et al. (2018). A novel vibration assisted polishing device based on the flexural mechanism driven by the piezoelectric actuators. *AIP Adv.* 8: 015012. https://doi.org/10.1063/1.5009027.

6 Kuhnen, K. (2003). Modeling, identification and compensation of complex hysteretic nonlinearities: a modified Prandtl-Ishlinskii approach. *Eur. J. Control.* https://doi.org/10.3166/ejc.9.407-418.

7 Moheimani, S.O.R. and Vautier, B.J.G. (2005). Resonant control of structural vibration using charge-driven piezoelectric actuators. *IEEE Trans. Control Syst. Technol.* https://doi.org/10.1109/TCST.2005.857407.

8 Al-Bender, F., Lampaert, V., and Swevers, J. (2005). The generalized Maxwell-slip model: a novel model for friction simulation and compensation. *IEEE Trans. Automat. Contr.* https://doi.org/10.1109/TAC.2005.858676.

9 Brehl, D.E. and Dow, T.A. (2008). Review of vibration-assisted machining. *Precis. Eng.* 32: 153–172. https://doi.org/10.1016/j.precisioneng.2007.08.003.

10 Kumar, S., Wu, C.S., Padhy, G.K., and Ding, W. (2017). Application of ultrasonic vibrations in welding and metal processing: a status review. *J. Manuf. Processes* 26: 295–322. https://doi.org/10.1016/j.jmapro.2017.02.027.

11 Amin, S.G., Ahmed, M.H.M., and Youssef, H.A. (1995). Computer-aided design of acoustic horns for ultrasonic machining using finite-element analysis. *J. Mater. Process. Tech.* 55: 254–260. https://doi.org/10.1016/0924-0136(95)02015-2.

12 Sinn, G., Zettl, B., Mayer, H., and Stanzl-Tschegg, S. (2005). Ultrasonic-assisted cutting of wood. *J. Mater. Process. Technol.* 170: 42–49. https://doi.org/10.1016/j.jmatprotec.2005.04.076.

13 Shen, X.H., Zhang, J.H., Yin, T.J., and Dong, C.J. (2010). A study on cutting force in micro end milling with ultrasonic vibration. *Adv. Mater. Res.* 97–101: 1910–1914. https://doi.org/10.4028/www.scientific.net/AMR.97-101.1910.

14 Adnan, A.S. and Subbiah, S. (2010). Experimental investigation of transverse vibration-assisted orthogonal cutting of AL-2024. *Int. J. Mach. Tools Manuf.* 50: 294–302. https://doi.org/10.1016/j.ijmachtools.2009.11.004.

15 Zhong, Z.W. and Lin, G. (2006). Ultrasonic assisted turning of an aluminium-based metal matrix composite reinforced with SiC particles. *Int. J. Adv. Manuf. Technol.* 27: 1077–1081. https://doi.org/10.1007/s00170-004-2320-3.

16 Shen, X.H., Zhang, J., Xing, D.X., and Zhao, Y. (2012). A study of surface roughness variation in ultrasonic vibration-assisted milling. *Int. J. Adv. Manuf. Technol.* 58: 553–561. https://doi.org/10.1007/s00170-011-3399-y.

17 Nath, C. and Rahman, M. (2008). Effect of machining parameters in ultrasonic vibration cutting. *Int. J. Mach. Tools Manuf.* 48: 965–974. https://doi.org/10.1016/j.ijmachtools.2008.01.013.

18 Liu, K., Li, X.P., Rahman, M., and Liu, X.D. (2004). A study of the cutting modes in the grooving of tungsten carbide. *Int. J. Adv. Manuf. Technol.* 24: 321–326. https://doi.org/10.1007/s00170-003-1565-6.

19 Liu, K., Li, X.P., Rahman, M., and Liu, X.D. (2004). Study of ductile mode cutting in grooving of tungsten carbide with and without ultrasonic vibration assistance. *Int. J. Adv. Manuf. Technol.* 24: 389–394. https://doi.org/10.1007/s00170-003-1647-5.

20 Ostasevicius, V., Gaidys, R., Dauksevicius, R., and Mikuckyte, S. (2013). Study of vibration milling for improving surface finish of difficult-to-cut materials. *Stroj. Vestnik/J. Mech. Eng.* 59: 351–357. https://doi.org/10.5545/sv-jme.2012.856.

21 Alam, K., Mitrofanov, A.V., and Silberschmidt, V.V. (2011). Experimental investigations of forces and torque in conventional and ultrasonically-assisted drilling of cortical bone. *Med. Eng. Phys.* 33: 234–239. https://doi.org/10.1016/j.medengphy.2010.10.003.

22 Kuo, K.L. (2008). Design of rotary ultrasonic milling tool using FEM simulation. *J. Mater. Process. Technol.* 201: 48–52. https://doi.org/10.1016/j.jmatprotec.2007.11.289.

23 Roy, S. (2017). Design of a circular hollow ultrasonic horn for USM using finite element analysis. *Int. J. Adv. Manuf. Technol.* 93: 319–328. https://doi.org/10.1007/s00170-016-8985-6.

24 Babitsky, V.I., Astashev, V.K., and Meadows, A. (2007). Vibration excitation and energy transfer during ultrasonically assisted drilling. *J. Sound Vib.* 308: 805–814. https://doi.org/10.1016/j.jsv.2007.03.064.

25 Hsu, C.Y., Huang, C.K., and Wu, C.Y. (2007). Milling of MAR-M247 nickel-based superalloy with high temperature and ultrasonic aiding. *Int. J. Adv. Manuf. Technol.* 34: 857–866. https://doi.org/10.1007/s00170-006-0657-5.

26 Moriwaki, T. and Shamoto, E. (1995). Ultrasonic elliptical vibration cutting. *CIRP Ann.* https://doi.org/10.1016/S0007-8506(07)62269-0.

27 Shamoto, E., Suzuki, N., Moriwaki, T., and Naoi, Y. (2002). Development of ultrasonic elliptical vibration controller for elliptical vibration cutting. *CIRP Ann. – Manuf. Technol.* 51: 327–330. https://doi.org/10.1016/S0007-8506(07)61528-5.

28 Shamoto, E. and Moriwaki, T. (1999). Ultraprecision diamond cutting of hardened steel by applying elliptical vibration cutting. *CIRP Ann. – Manuf. Technol.* 48: 441–444. https://doi.org/10.1016/S0007-8506(07)63222-3.

29 Liang, Z., Wang, X., Zhao, W. et al. (2010). A feasibility study on elliptical ultrasonic assisted grinding of sapphire substrate. *Int. J. Abras. Technol.* 3: 190–202. https://doi.org/10.1504/IJAT.2010.03405.

30 Liang, Z., Wang, X., Wu, Y. et al. (2013). Experimental study on brittle-ductile transition in elliptical ultrasonic assisted grinding (EUAG) of monocrystal sapphire using single diamond abrasive grain. *Int. J. Mach. Tools Manuf.* 71: 41–51. https://doi.org/10.1016/j.ijmachtools.2013.04.004.

31 Peng, Y., Liang, Z., Wu, Y. et al. (2012). Characteristics of chip generation by vertical elliptic ultrasonic vibration-assisted grinding of brittle materials. *Int. J. Adv. Manuf. Technol.* 62: 563–568. https://doi.org/10.1007/s00170-011-3839-8.

32 Liu, J., Zhang, D., Qin, L., and Yan, L. (2012). Feasibility study of the rotary ultrasonic elliptical machining of carbon fiber reinforced plastics (CFRP). *Int. J. Mach. Tools Manuf.* 53: 141–150. https://doi.org/10.1016/j.ijmachtools.2011.10.007.

33 Geng, D., Zhang, D., Xu, Y. et al. (2015). Rotary ultrasonic elliptical machining for side milling of CFRP: tool performance and surface integrity. *Ultrasonics* 59: 128–137. https://doi.org/10.1016/j.ultras.2015.02.006.

34 Suzuki, N., Yan, Z., Hino, R. et al. (2006). Ultraprecision micro-machining of single crystal germanium by applying elliptical vibration cutting. In: *2006 IEEE International Symposium on Micro-Nano Mechanical and Human Science*. MHS https://doi.org/10.1109/MHS.2006.320323.

35 Suzuki, N., Haritani, M., Yang, J. et al. (2007). Elliptical vibration cutting of tungsten alloy molds for optical glass parts. *CIRP Ann. – Manuf. Technol.* 56: 127–130. https://doi.org/10.1016/j.cirp.2007.05.032.

36 Suzuki, H., Hamada, S., Okino, T. et al. (2010). Ultraprecision finishing of micro-aspheric surface by ultrasonic two-axis vibration assisted polishing. *CIRP Ann. – Manuf. Technol.* 59: 347–350. https://doi.org/10.1016/j.cirp.2010.03.117.

37 Li, X. and Zhang, D. (2006). Ultrasonic elliptical vibration transducer driven by single actuator and its application in precision cutting. *J. Mater. Process. Technol.* 180: 91–95. https://doi.org/10.1016/j.jmatprotec.2006.05.007.

38 Yin, Z., Fu, Y., Li, H. et al. (2017). Mathematical modeling and experimental verification of a novel single-actuated ultrasonic elliptical vibrator. *Adv. Mech. Eng.* 9: 1–12. https://doi.org/10.1177/1687814017745413.

39 Yin, Z., Fu, Y., Xu, J. et al. (2017). A novel single driven ultrasonic elliptical vibration cutting device. *Int. J. Adv. Manuf. Technol.* 90: 3289–3300. https://doi.org/10.1007/s00170-016-9641-x.

40 Börner, R., Winkler, S., Junge, T. et al. (2018). Generation of functional surfaces by using a simulation tool for surface prediction and micro structuring of cold-working steel with ultrasonic vibration assisted face milling. *J. Mater. Process. Technol.* 255: 749–759. https://doi.org/10.1016/j.jmatprotec.2018.01.027.

41 Tan, R., Zhao, X., Zou, X., and Sun, T. (2018). A novel ultrasonic elliptical vibration cutting device based on a sandwiched and symmetrical structure. *Int. J. Adv. Manuf. Technol.* 97: 1397–1406. https://doi.org/10.1007/s00170-018-2015-9.

42 Guo, J., Suzuki, H., Morita, S.-Y. et al. (2012). Micro polishing of tungsten carbide using magnetostrictive vibrating polisher. *Key Eng. Mater.* 516: 569–574. https://doi.org/10.4028/www.scientific.net/KEM.516.569.

43 Guo, J., Morita, S.Y., Hara, M. et al. (2012). Ultra-precision finishing of micro-aspheric mold using a magnetostrictive vibrating polisher. *CIRP Ann. – Manuf. Technol.* 61: 371–374. https://doi.org/10.1016/j.cirp.2012.03.141.

44 Guo, J., Suzuki, H., and Higuchi, T. (2013). Development of micro polishing system using a magnetostrictive vibrating polisher. *Precis. Eng.* 37: 81–87. https://doi.org/10.1016/j.precisioneng.2012.07.003.

45 Asumi, K., Fukunaga, R., Fujimura, T., and Kurosawa, M.K. (2009). Miniaturization of a V-shape transducer ultrasonic motor. *Jpn. J. Appl. Phys.* https://doi.org/10.1143/JJAP.48.07GM02.

46 Zhang, F., Chen, W., Liu, J., and Wang, Z. (2005). Bidirectional linear ultrasonic motor using longitudinal vibrating transducers. *IEEE Trans. Ultrason. Ferroelectr. Freq. Control* https://doi.org/10.1109/TUFFC.2005.1397358.

47 Kurosawa, M.K., Kodaira, O., Tsuchitoi, Y., and Higuchi, T. (1998). Transducer for high speed and large thrust ultrasonic linear motor using two sandwich-type vibrators. *IEEE Trans. Ultrason. Ferroelectr. Freq. Control* https://doi.org/10.1109/58.726442.

48 Guo, P. and Ehmann, K.F. (2013). Development of a tertiary motion generator for elliptical vibration texturing. *Precis. Eng.* 37: 364–371. https://doi.org/10.1016/j.precisioneng.2012.10.005.

49 Song, D., Zhao, J., Ji, S., and Zhou, X. (2016). Development of a novel two-dimensional ultrasonically actuated polishing process. *AIP Adv.* 6 https://doi.org/10.1063/1.4967292.

50 Yanyan, Y., Bo, Z., and Junli, L. (2009). Ultraprecision surface finishing of nano-ZrO_2 ceramics using two-dimensional ultrasonic assisted grinding. *Int. J. Adv. Manuf. Technol.* 43: 462–467. https://doi.org/10.1007/s00170-008-1732-x.

51 Suzuki, N., Hino, R., and Shamoto, E. (2007). Development of 3 DOF ultrasonic elliptical vibration system for elliptical vibration cutting. In: *Proceedings of ASPE Spring Topical Meeting on Vibration Assisted Machining Technology*. ASPE. 40.

52 Shamoto, E., Suzuki, N., Tsuchiya, E. et al. (2005). Development of 3 DOF ultrasonic vibration tool for elliptical vibration cutting of sculptured surfaces. *CIRP Ann. – Manuf. Technol.* https://doi.org/10.1016/S0007-8506(07)60113-9.

53 Shamoto, E., Suzuki, N., Hino, R., et al. (2005). A new method to machine sculptured surfaces by applying ultrasonic elliptical vibration cutting. *Proceedings of the 2005 International Symposium on Micro-Nanomechatronics and Human Science, Eighth Symposium on* Micro- and Nano-Mechatronics for Information-based Society – The 21st Century COE Program. Nagoya University, 7 November 2005, Nagoya Municipal Industrial Research Institute, 8–9 November 2005. IEEE. https://doi.org/10.1109/MHS.2005.1589969.

54 Lee, D.Y., Kim, D.M., Gweon, D.G., and Park, J. (2007). A calibrated atomic force microscope using an orthogonal scanner and a calibrated laser interferometer. *Appl. Surf. Sci.* 253: 3945–3951. https://doi.org/10.1016/j.apsusc.2006.08.027.

55 Syahputra, H.P., Ko, T.J., and Chung, B.M. (2014). Development of 2-axis hybrid positioning system for precision contouring on micro-milling operation. *J. Mech. Sci. Technol.* https://doi.org/10.1007/s12206-013-1132-5.

56 Kurniawan, R., Ko, T.J., Ping, L.C. et al. (2017). Development of a two-frequency, elliptical-vibration texturing device for surface texturing. *J. Mech. Sci. Technol.* 31: 3465–3473. https://doi.org/10.1007/s12206-017-0635-x.

57 Hong, M.S. and Ehmann, K.F. (1995). Generation of engineered surfaces by the surface-shaping system. *Int. J. Mach. Tools Manuf.* 35: 1269–1290. https://doi.org/10.1016/0890-6955(94)00114-Y.

58 Greco, A., Raphaelson, S., Ehmann, K. et al. (2009). Surface texturing of tribological interfaces using the vibromechanical texturing method. *J. Manuf. Sci. Eng.* 131: 061005. https://doi.org/10.1115/1.4000418.

59 Suzuki, H., Marshall, M.B., Sims, N.D., and Dwyer-Joyce, R.S. (2017). Design and implementation of a non-resonant vibration-assisted machining device to create bespoke surface textures. *Proc. Inst. Mech. Eng. Part C J. Mech. Eng. Sci.* 231: 860–875. https://doi.org/10.1177/0954406215625087.

60 Wang, G. (2001). *Elliptical Diamond Milling: Kinematics, Force and Tool Wear.* North Carolina State University http://repository.lib.ncsu.edu/ir/handle/1840.16/2454.

61 Brehl, D.E., Dow, T., Garrard, K. et al. (2006). Micro-structure fabrication using elliptical vibration-assisted machining (Evam) 1. In: *Proceedings of the 21st Annual ASPE Meeting*, 511–514. ASPE.

62 Cerniway, M. (2001). *Elliptical Diamond Milling: Kinematics, Force and Tool Wear.* North Carolina State University.

63 Kim, G.D. and Loh, B.G. (2010). Machining of micro-channels and pyramid patterns using elliptical vibration cutting. *Int. J. Adv. Manuf. Technol.* 49: 961–968. https://doi.org/10.1007/s00170-009-2451-7.

64 Kim, G.D. and Loh, B.G. (2007). An ultrasonic elliptical vibration cutting device for micro V-groove machining: kinematical analysis and micro V-groove machining characteristics. *J. Mater. Process. Technol.* 190: 181–188. https://doi.org/10.1016/j.jmatprotec.2007.02.047.

65 Kim, G.D. and Loh, B.G. (2007). Characteristics of chip formation in micro V-grooving using elliptical vibration cutting. *J. Micromech. Microeng.* 17: 1458–1466. https://doi.org/10.1088/0960-1317/17/8/007.

66 Li, G., Wang, B., Xue, J. et al. (2018). Development of vibration-assisted micro-milling device and effect of vibration parameters on surface quality and exit-burr. *Proc. Inst. Mech. Eng. Part B J. Eng. Manuf.* https://doi.org/10.1177/0954405418774592.

67 Loh, B.G. and Kim, G.D. (2012). Correcting distortion and rotation direction of an elliptical trajectory in elliptical vibration cutting by modulating phase and relative magnitude of the sinusoidal excitation voltages. *Proc. Inst. Mech. Eng. Part B J. Eng. Manuf.* 226: 813–823. https://doi.org/10.1177/0954405411431375.

68 Ahn, J., Lim, H., and Son, S. (1999). Improvement of micro-machining accuracy by 2-dimensional vibration cutting. *Proc. ASPE* 20: 150–153. http://aspe.net:16080/publications/Annual_1999/POSTERS/PROCESS/NONCONV/AHN.PDF.

69 Chern, G.L. and Chang, Y.C. (2006). Using two-dimensional vibration cutting for micro-milling. *Int. J. Mach. Tools Manuf.* 46: 659–666. https://doi.org/10.1016/j.ijmachtools.2005.07.006.

70 Jin, X. and Xie, B. (2015). Experimental study on surface generation in vibration-assisted micro-milling of glass. *Int. J. Adv. Manuf. Technol.* 81: 507–512. https://doi.org/10.1007/s00170-015-7211-2.

71 Ding, H., Chen, S.J., and Cheng, K. (2010). Two-dimensional vibration-assisted micro end milling: cutting force modelling and machining process dynamics. *Proc. Inst. Mech. Eng. Part B J. Eng. Manuf.* 224: 1775–1783. https://doi.org/10.1243/09544054JEM1984.

72 Ding, H., Chen, S.J., and Cheng, K. (2011). Dynamic surface generation modeling of two-dimensional vibration-assisted micro-end-milling. *Int. J. Adv. Manuf. Technol.* 53: 1075–1079. https://doi.org/10.1007/s00170-010-2903-0.

73 Li, G. (2012). Non-resonant vibration auxiliary table development and study on micro-milling technology experiment. PhD thesis. Harbin Institute of Technology.

74 Zhang, J. and Sun, B. (2006). Design and analysis of 2-DOF nanopositioning stage based on dual flexure hinges. *Piezoelectr. Acoustoopt.* 28: 624–626.

75 Zheng, L., Chen, W., Pozzi, M. et al. (2018). Modulation of surface wettability by vibration assisted milling. *Precis. Eng.*: 1–10. https://doi.org/10.1016/j.precisioneng.2018.09.006.

76 Zheng, L., Chen, W., and Huo, D. (2018). Experimental investigation on burr formation in vibration-assisted micro-milling of Ti–6Al–4V. *Proc. Inst. Mech. Eng. Part C J. Mech. Eng. Sci.* https://doi.org/10.1177/0954406218792360.

77 Chen, W., Zheng, L., and Huo, D. (2018). Surface texture formation by non-resonant vibration assisted micro milling. *J. Micromech. Microeng.* 28: 025006.

78 Ibrahim, R. (2010). *Vibration Assisted Machining : Modelling, Simulation, Optimization, Control and Applications*. School of Engineering and Design Brunel University.

79 Ding, H., Chen, S.J., Ibrahim, R., and Cheng, K. (2011). Investigation of the size effect on burr formation in two-dimensional vibration-assisted micro end milling. *Proc. Inst. Mech. Eng. Part B J. Eng. Manuf.* 225: 2032–2039. https://doi.org/10.1177/0954405411400820.

80 Liu, Y. (2015). *The Non-resonant Elliptical Vibration Turning of Microstructure Surface*. Jilin University.

81 Kim, H.S., Kim, S.I., Il Lee, K. et al. (2009). Development of a programmable vibration cutting tool for diamond turning of hardened mold steels. *Int. J. Adv. Manuf. Technol.* 40: 26–40. https://doi.org/10.1007/s00170-007-1311-6.

82 Lin, J., Han, J., Lu, M. et al. (2017). Design, analysis and testing of a new piezoelectric tool actuator for elliptical vibration turning. *Smart Mater. Struct.* https://doi.org/10.1088/1361-665x/aa71f0.

83 Chee, S.K., Suzuki, H., Okada, M. et al. (2011). Precision polishing of micro mold by using piezoelectric actuator incorporated with mechanical amplitude magnified mechanism. *Adv. Mater. Res.* 325: 470–475. https://doi.org/10.4028/www.scientific.net/AMR.325.470.

84 Chee, S.K., Suzuki, H., Uehara, J. et al. (2013). A low contact force polishing system for micro molds that utilizes 2-dimensional low frequency vibrations (2DLFV) with piezoelectric actuators (PZT) and a mechanical transformer mechanism. *Int. J. Autom. Technol.* 7: 71–82.

85 Jieqiong, L., Yingchun, L., and Xiaoqin, Z. (2014). Tool path generation for fabricating optical freeform surfaces by non-resonant three-dimensional elliptical vibration cutting. *Proc. Inst. Mech. Eng. Part C J. Mech. Eng. Sci.* 228: 1208–1222. https://doi.org/10.1177/0954406213502448.

86 Li, Y. and Xu, Q. (2011). A totally decoupled piezo-driven XYZ flexure parallel micropositioning stage for micro/nanomanipulation. *IEEE Trans. Autom. Sci. Eng.* https://doi.org/10.1109/TASE.2010.2077675.

87 Wang, G. (2012). *Development of a Three-Dimensional Elliptical Vibration Assisted Diamond Cutting Apparatus*. Harbin Institute of Technology.

88 Liu, P. (2013). *Development of a New Apparatus for Three-Dimensional Elliptical Vibration Cutting*. Jilin University.

89 Zhu, Z., To, S., Ehmann, K.F., and Zhou, X. (2017). Design, analysis, and realization of a novel piezoelectrically actuated rotary spatial vibration system for micro-/nanomachining. *IEEE/ASME Trans. Mechatron.* 22: 1227–1237. https://doi.org/10.1109/TMECH.2017.2682983.

90 Lin, J., Lu, M., and Zhou, X. (2016). Development of a non-resonant 3D elliptical vibration cutting apparatus for diamond turning. *Exp. Tech.* 40: 173–183. https://doi.org/10.1007/s40799-016-0021-0.

91 Lu, M. (2014). *Development of 3D Elliptical Vibration Assisted Cutting Apparatus and its Control*. Jilin University.

92 Zhang, Y., Kang, R., Liu, J. et al. (2017). Review of ultrasonic vibration assisted drilling. *J. Mech. Eng.* 53 https://doi.org/10.3901/JME.2017.19.033.

3

Vibration System Design and Implementation

3.1 Introduction

According to operating frequency, vibration devices can be divided into two groups: resonant mode systems and nonresonant mode systems. For a resonant system, a sonotrode (also called a horn or a concentrator) vibrates at its natural frequency, transferring, and amplifying a given vibration from a vibration source, which is usually a magnetostrictive or piezoelectric transducer. It can achieve higher operating frequency and greater energy efficiency compared with nonresonant systems. However, its vibration trajectory cannot be controlled precisely owing to the nature of resonant vibrations and the phase lag between excitation and the mechanical response. A nonresonant vibration device applies forced vibration theory and can produce variable vibration frequencies, which are always less than its natural frequency. Compared with resonant systems, nonresonant systems tend to achieve higher vibration accuracy, and it is easier to achieve closed loop control of the vibration trajectories under low-frequency conditions. As one of the most promising ultraprecision motion mechanisms, a flexible mechanism driven by a piezoelectric actuator is widely applied in ultraprecision manufacturing for the optical, biomedical, and aerospace industries because of its merits such as rapid response, high movement resolution, no friction and wear, and compact structure, and also, no lubrication is required. Nonresonant vibration-assisted machining with the working frequency less than the device's natural frequency makes precise closed loop control possible and has received increasing attention. However, the full decoupling mechanism increases its design complexity and leads to some sacrifices in performance. In addition, the piezoelectric actuator can only withstand compressive stress, whereas the coupling motion mechanism generates shear stress and can cause damage.

Currently, serial and parallel types of flexible structure have been developed to reduce the coupling effect. The features of the serial type flexible structure include ease of control and low motion coupling, but it also has drawbacks. The complex structure leads to low structural stiffness, working bandwidth, and response rate, and motion errors at various levels in the kinematic chain are transmitted step by step so that a large error accumulates during operation. Furthermore, it is difficult to achieve consistent dynamic performance for each axis of motion. On the other hand, the benefit of the parallel type of flexible structure is

Vibration Assisted Machining: Theory, Modelling and Applications,
First Edition. Lu Zheng, Wanqun Chen, and Dehong Huo.

its compact structure, which leads to high stiffness, bandwidth, and motion accuracy. However, the coupling effect between different motion axes cannot effectively be suppressed and a double symmetrical structure is always required. In this chapter, a general understanding of vibration system design from the beginning to implementation is given, including for both resonant and nonresonant systems.

3.2 Resonant Vibration System Design

3.2.1 Composition of the Resonance System and Its Working Principle

The ultrasonic machining system is mainly composed of two sections: a precision machine tool and an ultrasonic vibration device. The ultrasonic vibration device can be divided into three parts: an ultrasonic generator, a transducer, and a horn.

The ultrasonic generator is mainly constituted by an oscillator, a voltage amplifier, a power amplifier, and an output transformer. Among these, the oscillator is the core of the ultrasonic frequency Both sinusoidal and non-sinusoidal output waveforms can be generated to satisfy the machining demand. But the sinusoidal output waveform is the most common type.

The ultrasonic transducer is an energy conversion device that converts an alternating electrical signal into the acoustic signal or converts an acoustic signal in an external sound field into an electrical signal in the ultrasonic frequency range. The most commonly used are hysteresis and piezoelectric transducers.

The ultrasonic horn is an important component that transmits the mechanical energy converted from electrical energy to the workpiece. It involves a mechanical amplification of the power ultrasonic amplitude so as to improve ultrasonic processing efficiency.

The basic principle of the ultrasonic vibration system is that the ultrasonic generator converts 220 or 380 V AC into an ultrasonic frequency oscillation signal with a certain power output and transmits it to the transducer, usually using piezoelectric ceramics. Then, the transducer converts the high-frequency oscillating electrical signal into a high-frequency mechanical vibration signal, and this is amplified by the horn and finally transmitted to the workpiece or the tool.

3.2.2 Summary of Design Steps

In the design stage, the diameter and number of piezoelectric plates, as well as the shape of front and rear metal cover plates, need to be determined by considering the application environment and the power supply. For ultrasonic transducer design, the plane with zero particles' displacement is the nodal plane, and this needs to be confirmed before calculating the size of the cover plates. The nodal plane can be set in different positions, including at the interface between the rear cover and the ceramic stack, the middle of the ceramic stack, or the interface between the ceramic stack and the rear cover. The size of the front and rear metal cover plates can be obtained by calculating its resonance conditions without taking into account the effect of machining load, which allows the calculation to be made for no-load transducer conditions where it is free at both ends.

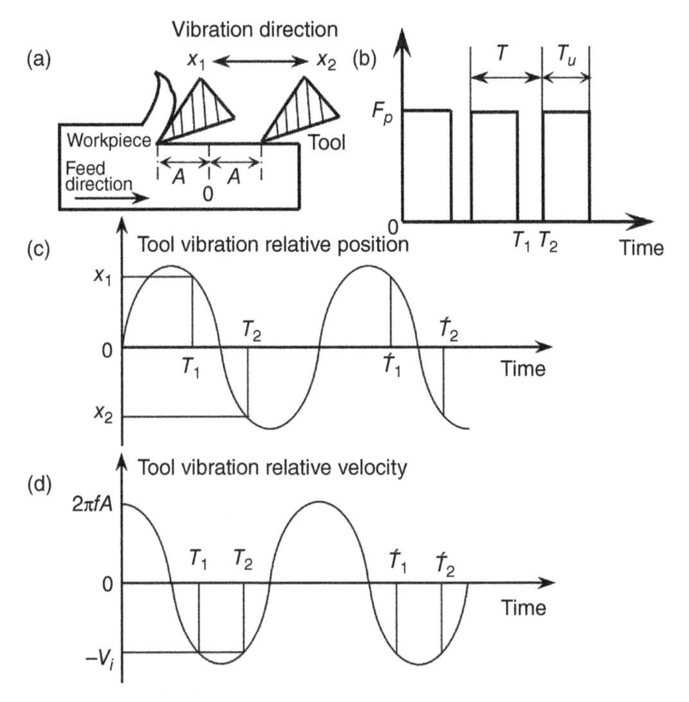

Figure 3.1 Schematic of vibration-assisted machining: (a) relative movement of workpiece and tool; (b) cutting force; (c) tool vibration relative position; and (d) tool vibration relative velocity.

3.2.3 Power Calculation

The power of the vibrator can be determined based on the displacement of the piezoceramics required and the selected piezoceramic model, as well as the input voltage. Ultrasonic excitation sources with large power margins are often used because of the lack of a correct understanding of the ultrasonic vibration cutting energy and power. The power of the excitation source should be determined in two respects: one is the energy required to complete the vibration cutting itself and the other is the energy transmission efficiency that includes the efficiency of converting electrical energy into mechanical energy and the transmission efficiency of mechanical energy. Because the excitation source power required for ultrasonic vibration cutting is not itself large, the key to a good vibration cutting system is to select a highly efficient transducer and avoid the transmission loss of mechanical energy as far as possible.

Figure 3.1 shows the forces from the tool during the vibration-assisted machining process. The pulse force of the cutting tool is F_p and only during T_u time is the workpiece in contact with the tool working in each cycle of the cutting process T. At that time, the cutting force is F_p and the working length is L_{pu}. Therefore, the power P_v required in the vibration-assisted machining process is

$$P_v = (F_p * L_{pu}) * f \tag{3.1}$$

where f is the vibration frequency.

3.2.3.1 Analysis of Working Length L_{pu}

The kinematics of the tool needs to be understood first in order to analyze the working length L_{pu}. Figure 3.1 shows the simplified trajectory of the tool tip, which is harmonically set in motion at frequency f and amplitude A. The tool-relative position and velocity are given by

$$x(t) = Vt + A \sin 2\pi ft \tag{3.2}$$

$$\dot{x}(t) = V + 2\pi fA \cos 2\pi ft \tag{3.3}$$

where t is the cutting time.

Equation (3.3) leads to the definition of the critical velocity V_c, above which the tool never separates from the uncut workpiece:

$$V_c = 2\pi fA \tag{3.4}$$

If the cutting speed V is less than V_c, periodic interruption of cutting takes place at a frequency of f. If the cutting speed V is larger than V_c, then cutting is continuous, although the relative velocity between the tool and workpiece varies harmonically. In addition, this case only considers the cutting condition where the cutting speed is less than the critical velocity V_c.

T_1 is set to the time from the start of cutting to the cutting edge leaving the chip. When the tool and chip have disengaged, the cutting speed V_i is exactly equal to the tip vibration speed $\dot{x}$, that is:

$$\dot{x} = 2A\pi f \cos 2\pi fT_1 = -V_i \tag{3.5}$$

$$T_1 = \frac{\cos^{-1}\left(-\dfrac{V}{2a\pi f}\right)}{2\pi f} \tag{3.6}$$

Set the time of next contact between tool tip and workpiece at T_2, that is:

$$x_1 - x_2 = V(T_1 - T_2) \tag{3.7}$$

where x_1 and x_2 are the displacements of the tool tip at times T_1 and T_2, respectively, and Eq. (3.7) can be expressed as:

$$x_1 = A \sin 2\pi fT_1 \tag{3.8}$$

$$x_2 = A \sin 2\pi fT_2 \tag{3.9}$$

$$T_1 - T_2 = \frac{A \sin 2\pi fT_1 - A \sin 2\pi fT_2}{V} \tag{3.10}$$

Therefore, the effective cutting time T_u is:

$$T_u = T + T_1 - T_2 \tag{3.11}$$

In addition, the working length L_{pu} can be expressed as:

$$L_{pu} = \int_0^{T_1} (\dot{x} + V)dt + \int_{T_2}^{T} (\dot{x} + V)dt \tag{3.12}$$

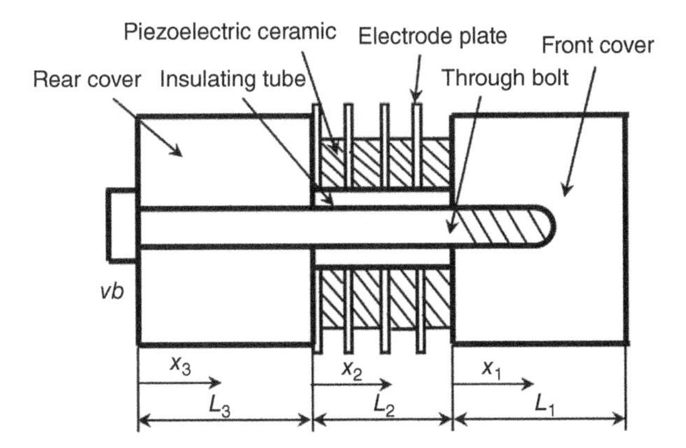

Figure 3.2 Schematic diagram of an ultrasonic transducer.

3.2.3.2 Analysis of Cutting Tool Pulse Force F_p

In the vibration-assisted machining process, the impact load acts between the cutting tool and the workpiece, which improves the cutting performance. Meanwhile, impact damage to the workpiece consumes less cutting energy and reduces the cutting force. Moreover, internal friction between the rake face and the shear face of the tool converts external friction and leads to lower friction, which further reduces the cutting force. Therefore, both the average cutting force and pulse force are reduced in vibration-assisted machining, and according to experiment results, the pulse force F_p represents almost 70% of the cutting force in conventional machining.

3.2.3.3 Calculation of Total Required Power

The power of the piezoelectric ceramic consists of two parts: the cutting power (P_r) and the other unloaded power (P_u). The unload power of the vibrator (P_u) can be expressed as:

$$P_u = K_p * A * A * f \tag{3.13}$$

where K_p is the stiffness of the piezoceramic, A is the required vibration amplitude (assume 2 μm), and f is the required vibration frequency (assume 30 kHz).

Therefore, the total required power is:

$$P_t = P_u + P_v \tag{3.14}$$

3.2.4 Ultrasonic Transducer Design

Figure 3.2 shows a typical schematic diagram of an ultrasonic transducer, which mainly includes rear and front cover plates, the piezoelectric ceramic plate, electrode plate, and a through bolt. Among these parts, the selection of the piezoelectric ceramic plate and the rear and front cover plate design are the most important factors and have a great influence on transducer performance.

3.2.4.1 Piezoelectric Ceramic Selection

The size of the piezoelectric ceramic component mainly refers to the geometrical dimension of the single circular ring of the piezoelectric ceramic in the vibration direction and the total volume of the entire piezoelectric ceramic crystal stack. The achievement of maximum efficiency with a minimum volume and weight is always desired, which is closely related to the material and type of piezoelectric ceramics involved. For a longitudinal vibration transducer, the antiresonance frequency of the piezoelectric ceramic's thin disc thickness in stretching vibration mode is inversely proportional to the thickness, that is:

$$f_a = \frac{n}{2t}\sqrt{\frac{c_{33}^D}{\rho}}, \quad n = 1, 3, 5\ldots \tag{3.15}$$

where f_a is the resonant frequency, n is the resonant order, c_{33}^D is the elastic stiffness constant, t is the thickness of the piezoelectric ceramic plate, and ρ is the density of the piezoelectric ceramic plate.

In addition, a high-bandwidth vibration frequency, from 3 MHz to dozens of MHz, can be obtained in the thickness stretching vibration mode because of the thin nature of the piezoelectric ceramic plate.

By analyzing the vibration mode of the piezoelectric ceramic vibrator, the piezoelectric ceramic plate or ring in longitudinal and thickness vibration modes has a relatively high electromechanical coupling coefficient. Therefore, the axially polarized piezoelectric ceramic plate or ring is the best choice for the transducer, and a high electroacoustic conversion efficiency at the ultrasound power level can be obtained. However, for a piezoelectric ceramic vibrator, the thickness in the polarization direction should be 4 cm or more in order to obtain a resonance frequency of 50 kHz or less. It is quite difficult to sinter and polarize because of high internal impedance. In order to overcome this difficulty, a sandwich-type piezoelectric ceramic transducer composed of metal blocks on both end faces of a piezoelectric ceramic wafer is often used. Since this structure of transducer was first proposed by the French physicist Langevin, it is also called a Langevin transducer. The direction of polarization in the piezoelectric ceramic ring in this type of transducer is consistent with the thickness of the vibrating part. The piezoelectric ceramic ring is connected to the metal blocks at both ends by a stress through bolt whose thickness is equal to a half wavelength of the fundamental vibration wave. The features of this structure of the transducer are such that the desired resonant frequency can be generated using the longitudinal inverse piezoelectric effect of the piezoelectric ceramic ring set. In addition, the piezoelectric ceramic ring is prone to cracking during operation because of its poor tensile strength (high compressive strength). The piezoelectric ceramic ring of a Langevin transducer is preloaded by a combination of the metal front and rear cover plates and the through bolt, which indicates that it is always in a compressed state under strong vibration to avoid its cracking.

For the longitudinal vibration transducer, the d33 excitation vibration mode is usually used for the piezoelectric ceramic plate in the thickness direction, which requires a piezoelectric material with a high piezoelectric constant and electromechanical coupling coefficient k33 as well as low dielectric loss and good mechanical properties. After comparing the properties of different piezoelectric materials, the PZT-8 emission type of the piezoelectric ceramic material with a high comprehensive index is selected.

In the longitudinal vibration transducer design, the diameter of the piezoelectric ceramic ring should be as far as possible less than one quarter of the wavelength of the acoustic wave in the ceramic material. Therefore, its limiting outer diameter D can be expressed as:

$$D \leq \frac{\lambda}{4} = \frac{c}{4f} \tag{3.16}$$

where λ is the wavelength of the acoustic wave in the ceramic material, c is the wave speed of the acoustic wave in the ceramic material, and f is the frequency of the acoustic wave in the ceramic material.

For the piezoelectric PZT-8, the wave speed of the acoustic wave in its ceramic material is 3.1×10^5 cm/s, and the limit outer diameter D can be calculated, which is 38.8 mm. During the design process of piezoelectric ceramic, the actual outer diameter can be slightly larger than the limit outer diameter because of its ring structure. Recommended parameters of the piezoelectric ceramic are outer diameter $D = 50$ mm, inner diameter $d = 20$ mm, and thickness $h = 5$ mm. Therefore, the number of piezoelectric ceramic rings n required by the transducer can be expressed as:

$$n \approx \frac{P}{\frac{\pi}{4} \times (D^2 - d^2) \times h \times f \times P_c} \tag{3.17}$$

where P is the power of the transducer and can be assumed as 900 W, P_c is the power capacity of the PZT material, which is $2\,\mathrm{W/cm^3}$, and f is the resonant frequency and can be assumed as 3×10^4 Hz.

In order to connect the front and rear covers of the transducer to the electrodes at the same polarity, the number of piezoelectric ceramic rings is usually an even number. In this design, four piezoelectric ceramic rings ($n = 4$) are used, which require four copper electrodes with a thickness of 0.3 mm. Then, the transducer length (l_2) is:

$$l_2 = 5 * 4 + (4 * 0.3) = 21.2\ \mathrm{mm} \tag{3.18}$$

3.2.4.2 Calculation of Back Cover Size

For a Langevin transducer, high-density metals such as steel or copper are always selected for back cover materials. Meanwhile, low-density metals with good sound characteristics and heat dissipation performance such as aluminum or titanium alloys are always chosen for the front cover metal. This can reduce the overall weight, obtain a larger displacement ratio of the front and rear cover plates, improve the forward radiation capability of the transducer, and enable it to achieve unobstructed one-way radiation.

3.2.4.3 Variable Cross-Sectional, One-Dimensional Longitudinal Vibration Wave Equation

The Langevin transducer shown in Figure 3.2 can be simplified as a variable cross-sectional bar, and it is assumed that this bar is made of a homogeneous, isotropic material, and its mechanical energy loss is ignored. Also, it is assumed that the cross-sectional area of the rod is much smaller than the wavelength of the vibration wave propagating along its axial direction and its stress distribution is uniform. Therefore, the wave equation of the one-dimensional longitudinal vibration of the variable cross-sectional rod can be expressed as:

$$\frac{\partial^2 \varepsilon}{\partial x^2} + \frac{1}{S(x)} \frac{\partial S(x)}{\partial x} \frac{\partial \varepsilon}{\partial x} + k^2 \varepsilon = 0 \tag{3.19}$$

where $S(x)$ represents the arbitrary cross-sectional area function, ε is the particle displacement function, k is the wavenumber, $k = \omega/c$, and ω is the vibration angle frequency, $\omega = 2\pi f$.

Generally, the arbitrary cross-sectional area function of the transducer is constant. Therefore, the wave velocity equation of each part of the transducer can be obtained from the Eq. (3.16) and expressed as:

$$\frac{\partial^2 v_n}{\partial x^2} + k_n^2 v_n = 0 \tag{3.20}$$

$$F_n(x_n) = \frac{ES_n(x)}{j\omega} \times \frac{\partial v_n}{\partial x_n} \tag{3.21}$$

where F is the elastic force, E is Young's modulus of elasticity, and x_n is the length dimension variable, $n = 1,2,3,4 \ldots$

Solving differential equations for the above formula gives the following:

$$v_n(x_n) = A_n \sin(k_n x_n) + B_n \cos(k_n x_n) \tag{3.22}$$

$$F_n(x_n) = (-j)\rho_n c_n s_n(x_n)[A_n \cos(k_n x_n) - B_n \sin(k_n x_n)] \tag{3.23}$$

The surface at which the piezoelectric ceramic ring is in contact with the horn is set as a nodal plane. Its boundary condition is as follows: $v_1(l_1) = 0$, $v_1(0) = v_2(l_2)$, $v_2(0) = v_3(l_3)$, $v_3(0) = v_b$, $F_1(0) = F_2(l_2)$, $F_2(0) = F_3(l_3)$, and $F_3(0) = 0$, and the following results can be obtained:

$$A_1 = -\frac{\cos(k_1 l_1)}{\sin(k_1 l_1)} V_b \left(-\frac{Z_3}{Z_2} \sin(k_2 l_2) \sin(k_3 l_3) + \cos(k_2 l_2) \cos(k_3 l_3)\right) \tag{3.24}$$

$$B_1 = V_b \left(-\frac{Z_3}{Z_2} \sin(k_2 l_2) \sin(k_3 l_3) + \cos(k_2 l_2) \cos(k_3 l_3)\right) \tag{3.25}$$

$$A_2 = -\frac{Z_3}{Z_2} V_b \sin(k_3 l_3) \tag{3.26}$$

$$B_2 = V_b \cos(k_3 l_3) \tag{3.27}$$

$$A_3 = 0 \tag{3.28}$$

$$B_3 = V_b \tag{3.29}$$

The frequency equation can be expressed as:

$$\frac{Z_3}{Z_2} \tan(k_2 l_2) \tan(k_3 l_3) + \frac{Z_3}{Z_1} \tan(k_1 l_1) \tan(k_3 l_3) + \frac{Z_2}{Z_1} \tan(k_2 l_2) \tan(k_1 l_1) = 0 \tag{3.30}$$

where the characteristic impedance of the transducer $Z_n = \rho_n c_n S_n(x_n)$.

If the nodal plane is taken to be the right-hand surface of the piezoelectric ceramic ring stack ($l_1 = 0$), then the left-hand quarter-wave oscillator part of the nodal plane is composed of only two parts, which are the piezoelectric ceramic and metal back cover, and Eq. (3.27) can be simplified as:

$$\frac{Z_3}{Z_2} \tan(k_2 l_2) \tan(k_3 l_3) = 1 \tag{3.31}$$

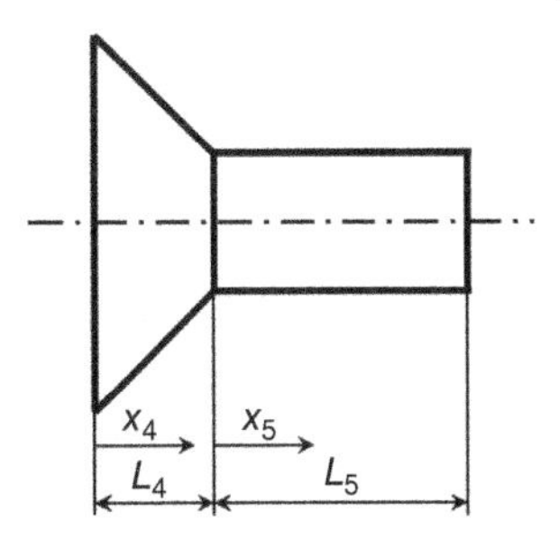

Figure 3.3 Layout of the quarter-wave horn.

3.2.4.4 Calculation of Size of Longitudinal Vibration Transducer Structure

Set transducer resonant frequency $f = 3 \times 10^4$ Hz and the electromechanical coupling coefficient of the piezoelectric ceramic rings (PZT-8) $K_{33} = 0.64$, the density of the piezoelectric ceramic rings $\rho_4 = 7.5\,\text{g/cm}^3$, sound speed in the piezoelectric ceramic rings $c_4 = 3.1 \times 10^5$ cm/s, copper electrode density $\rho = 8.9\,\text{g/cm}^3$, and the copper electrode sound speed $c = 3.1 \times 10^5$ cm/s. Carbide steel is chosen as the rear cover material and its density and sound speed are $\rho_5 = 7.8\,\text{g/cm}^3$ and $c_5 = 5.2 \times 10^5$ cm/s, respectively. Substituting l_2 into the frequency equation and the length of the rear cover can be obtained as:

$$l_3 = 4.8\,\text{mm} \tag{3.32}$$

A length of 5 mm is chosen for easy processing.

3.2.5 Horn Design

Quarter-wave horns are commonly used when designing composite horns or combination transducers. Generally, two specific locations at the big and the small ends of the variable section rod can be selected as the node points. If one end of the quarter-wave horn is set at the node point, the vibration displacement or velocity at that end is zero. In an ideal state, the amplification factor M_p and the input impedance Z_I at the node point are infinite for a lossless quarter-wave horn. However, material losses and the load at the other end of the rod cannot be ignored in practical applications, and so, both M_p and Z_I are limited. Because the quarter-wave horn features large values of M_p and Z_I, it is usually used as an impedance matching combination in the transducer design so as to improve the radiation efficiency of the transducer. The length of the horn can be calculated by using the two end areas and the vibration frequency.

Figure 3.3 shows a typical structure of a conical quarter-wave horn, which is always applied in conditions of a low amplification factor. Its frequency equation can be expressed as:

$$\tan(k_5 l_5) = \text{ctg}(k_4 l_4) + \frac{1}{k_4 l_4} \times \frac{R_4 - R_5}{R_5} \tag{3.33}$$

The amplification factor is:

$$\left|\frac{V_f}{V_b}\right| = \frac{R_3}{R_4} \cdot \frac{Z_1}{Z_3} \cdot \frac{\cos(k_2 l_2)}{\sin(k_1 l_1)} \cdot \frac{\sin(k_3 l_3)}{\cos(k_4 l_4)} \tag{3.34}$$

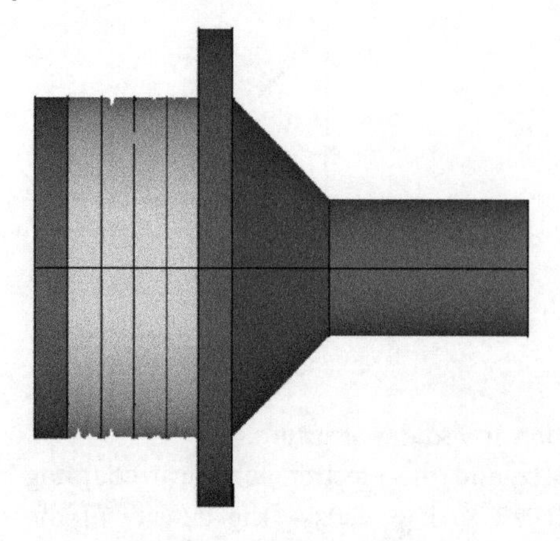

Figure 3.4 Finite element analysis mode of the ultrasonic vibrator.

In addition, the vibration speed ratio of the horn is:

$$\left|\frac{V_f}{V_b}\right| = \frac{R_4}{R_5} \cdot \frac{Z_2}{Z_4} \cdot \frac{\cos(k_3 l_3)}{\sin(k_2 l_2)} \cdot \frac{\sin(k_4 l_4)}{\cos(k_5 l_5)} \tag{3.35}$$

The large section length and wave number product are set to 0.3, the big end radius to 25 mm, and the small end radius to 10 mm. The length of the large and small sections can be calculated, which are $l_4 = 8$ mm and $l_5 = 40$ mm, respectively.

3.2.6 Design Optimization

According to the above parameters, the finite element model is developed and shown in Figure 3.4. The complete simulation program can be found in the appendix. The calculation results show that the first-order longitudinal vibration frequency of the vibrator is about 26 kHz, which has a large error compared with the design frequency of 30 kHz. This could due to many reasons:

(1) *Different boundary conditions.* The boundary conditions in the numerical calculation are free at both ends, while the finite element calculation uses fixed constraints at the flange.
(2) Numerical calculation does not consider the influence of the flange.

The optimized existing parameters are $l_4 = 12$ mm and $l_5 = 36$ mm. The first-order longitudinal mode changes to 30 940 Hz, as shown in Figure 3.5.

When an excitation signal with amplitude 2 μm and frequency 10 ~ 35 kHz is applied to the left-hand side of the horn, the displacement response of the small end of the horn is as shown in Figure 3.5.

No scrambling signal can be found in Figure 3.6 in the working range of the horn, and it can achieve accurate displacement output. The theoretical maximum resonance value is 50 μm, and the amplification ratio is 25 times. However, in fact, the theoretical resonance position is quite hard to achieve, and so the real value will be less than the simulated amplitude value.

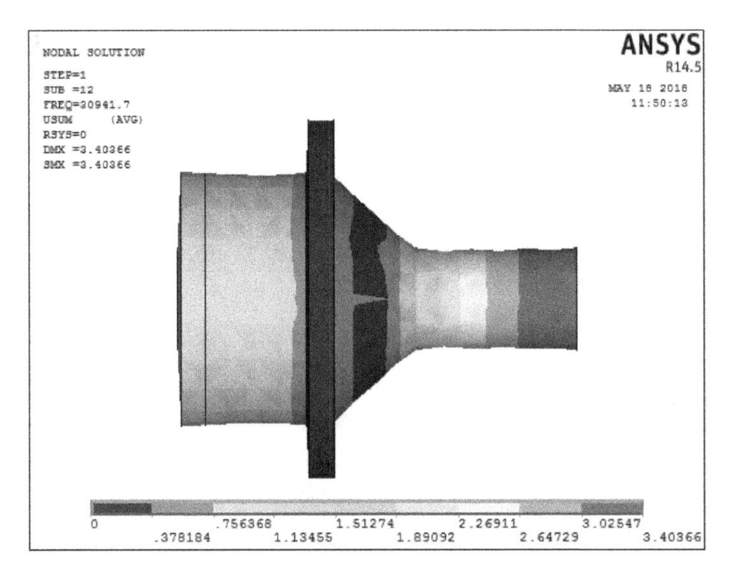

Figure 3.5 First-order longitudinal mode.

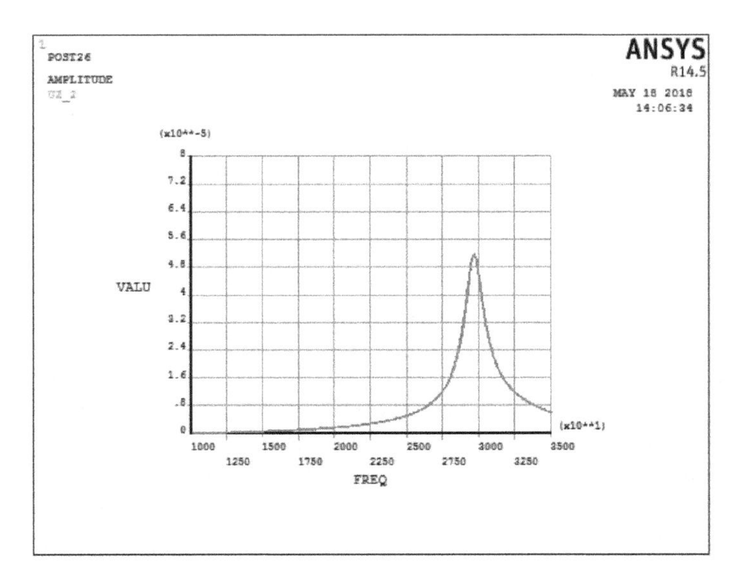

Figure 3.6 Displacement response signal at the small end of the horn.

3.3 Nonresonant Vibration System Design

As reviewed in Chapter 2, the design of the nonresonant system is quite different from that of the resonant system, and most designs work in two dimensions. Among these designs, double-parallel four-bar linkages with a flexure hinge structure are always used because of its high movement resolution and accuracy, as well as low coupling displacement. Figure 3.7 shows a two-dimensional nonresonant vibration stage, which has been

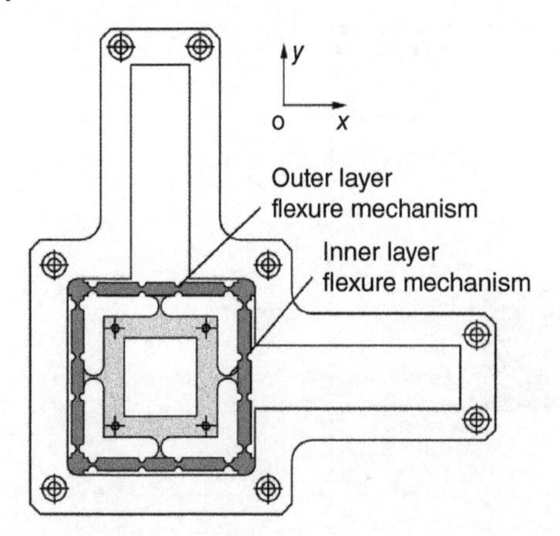

Figure 3.7 Layout of the designed vibration stage. Sources: Zheng et al. and Chen et al. [11–13].

reviewed in Chapter 2. In this section, the nonresonant vibration system design will be introduced for this vibration stage.

3.3.1 Modeling of Compliant Mechanism

Various methods have been developed for the modeling of compliant mechanisms, including pseudo-rigid body (PRB) method, Castigliano's second theorem, and inverse kinematic modeling (IKM). The pseudo-rigid body method equivalents the elastic rod (or hinge) model to the corresponding rigid rod model to achieve model simplification and is more suitable for planar structure [1, 2]. Castigliano's second theorem expresses the strain energy of the flexure structure as a function of the load. However, the calculation is quite complicated and involves partial differential equations, which is not conducive to the design optimization of the mechanism. Inverse kinematic modeling is the most accurate method compared with the other two methods. However, the computation can be time consuming. To overcome these drawbacks, a method named matrix-based compliance modeling (MCM) has been developed building upon the linear Hooke's law for the material [3]. The working principles of IKM and MCM are somewhat similar. They all apply the flexibility matrix of the flexure hinge unit and spatial coordinate transformation to build the stiffness/compliance model for a complex flexure mechanism. However, compared with the IKM method, the MCM replaces the Lagrange equation with the position transformation matrix, which improves computational efficiency and modeling accuracy.

3.3.2 Compliance Modeling of Flexure Hinges Based on the Matrix Method

In the modeling, each flexure hinge in the stage is uniformly divided into N pieces, and each of them can be treated as a micro-Euler–Bernoulli beam. This is more suitable for modeling spatial compliant mechanisms with complex structures and shapes because higher calculation accuracy and efficiency can easily be obtained using matrix operations. Therefore, the matrix-based method is chosen for the modeling of proposed 2D vibration stage.

The basic structural element in the 2D vibration stage is the right circular flexure hinge, which is shown in Figure 3.8, with the local Cartesian coordinates o_r–$x_r y_r z_r$. Taking into

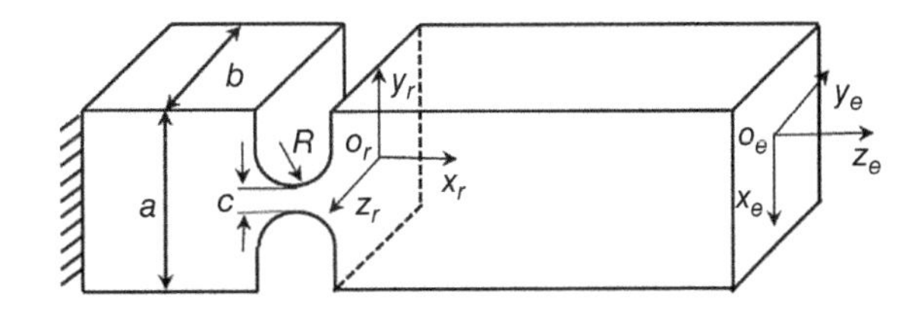

Figure 3.8 Schematic of flexure hinge.

account the complicated shear and torsion effects, a 6 × 6 matrix for describing the compliance C_i^r of every microbeam at o_r–$x_r y_r z_r$ is proposed, and this can be expressed as:

$$C_i^r = \begin{pmatrix} C_{1,1} & 0 & 0 & 0 & 0 & 0 \\ 0 & C_{2,2} & 0 & 0 & 0 & C_{2,6} \\ 0 & 0 & C_{3,3} & 0 & C_{3,5} & 0 \\ 0 & 0 & 0 & C_{\theta_x M_x} & 0 & 0 \\ 0 & 0 & C_{5,3} & 0 & C_{5,5} & 0 \\ 0 & C_{6,2} & 0 & 0 & 0 & C_{6,6} \end{pmatrix} \tag{3.36}$$

Set $s = c/R$, and G and E are the modulus of elasticity and rigidity, respectively. Moreover, v is Poisson's ratio, and the value of the matrix can be expressed as:

$$C_{1,1} = \frac{\delta x}{F_x} = \frac{1}{Eb}\left[\frac{2(2s+1)}{\sqrt{4s+1}} \arctan\sqrt{4s+1} - \frac{\pi}{2}\right] \tag{3.37}$$

$$C_{2,2} = \frac{\delta y}{F_y} = \left(\frac{\delta y}{F_y}\right) + \left(\frac{\delta y}{F_y}\right)_\tau \tag{3.38}$$

$$\left(\frac{\delta y}{F_y}\right) = \frac{12}{Eb}\left[\frac{s(24s^4+24s^3+22s^2+8s+1)}{2(2s+1)(4s+1)^2} + \frac{(2s+1)(24s^4+8s^3-14s^2-8s-1)}{2(4s+1)^{\frac{2}{5}}} \arctan\sqrt{4s+1} + \frac{\pi}{8}\right] \tag{3.39}$$

$$\left(\frac{\delta y}{F_y}\right)_\tau = \frac{1}{Gb}\left[\frac{2(2s+1)}{\sqrt{4s+1}} \arctan\sqrt{4s+1} - \frac{\pi}{2}\right] \tag{3.40}$$

$$C_{2,6} = \frac{\delta y}{M_z} = -\frac{12}{EbR}\left[\frac{2s^3(6s^2+4s+1)}{(2s+1)(4s+1)^2} + \frac{12s^4(2s+1)}{(4s+1)^{\frac{2}{5}}} \arctan\sqrt{4s+1}\right] \tag{3.41}$$

$$C_{3,3} = \frac{\delta z}{F_z} = \left(\frac{\delta z}{F_z}\right) + \left(\frac{\delta z}{F_z}\right)_\tau \tag{3.42}$$

$$\left(\frac{\delta z}{F_z}\right) = \frac{12R^2}{Eb^3}\left[\frac{2s+1}{2s} + \frac{(2s+1)(4s^2-4s-1)}{2s^2\sqrt{4s+1}} \arctan\sqrt{4s+1} - \frac{(2s^2-4s-1)\pi}{8s^2}\right] \tag{3.43}$$

$$\left(\frac{\delta z}{F_z}\right)_\tau = \frac{1}{Gb}\left[\frac{2(2s+1)}{\sqrt{4s+1}}\arctan\sqrt{4s+1} - \frac{\pi}{2}\right] \tag{3.44}$$

$$C_{3,5} = \frac{\delta z}{M_y} = \frac{12R}{Eb^3}\left[\frac{2(2s+1)}{\sqrt{4s+1}}\arctan\sqrt{4s+1} - \frac{\pi}{2}\right] \tag{3.45}$$

$$C_{5,3} = \frac{\delta\theta_y}{F_z} = \frac{R\delta\theta_y}{M_z} = \frac{12R}{Eb^3}\left[\frac{2(2s+1)}{\sqrt{4s+1}}\arctan\sqrt{4s+1} - \frac{\pi}{2}\right] \tag{3.46}$$

$$C_{5,5} = \frac{\delta\theta_y}{M_y} = \frac{12}{Eb^3}\left[\frac{2(2s+1)}{\sqrt{4s+1}}\arctan\sqrt{4s+1} - \frac{\pi}{2}\right] \tag{3.47}$$

$$C_{6,2} = \frac{\delta\theta_z}{F_y} = -\frac{R\delta\theta_y}{M_z} = -\frac{12}{EbR}\left[\frac{2s^3(6s^2+4s+1)}{(2s+1)(4s+1)^2} + \frac{12s^4(2s+1)}{(4s+1)^{\frac{5}{2}}}\arctan\sqrt{4s+1}\right] \tag{3.48}$$

$$C_{6,6} = \frac{\delta\theta_z}{M_z} = \frac{12}{EbR^2}\left[\frac{2s^3(6s^2+4s+1)}{(2s+1)(4s+1)^2} + \frac{12s^4(2s+1)}{(4s+1)^{\frac{5}{2}}}\arctan\sqrt{4s+1}\right] \tag{3.49}$$

where $\left(\frac{\delta}{F}\right)_\tau$ is the linear deformation of the flexure hinge caused by the shear force under the action of the load F.

By arranging the above equations, a widely accepted flexible matrix is represented as follows [4, 5]:

$$C_i^r = \begin{pmatrix} \frac{L}{Ea_ib_i} & 0 & 0 & 0 & 0 & 0 \\ 0 & \frac{4L^3}{Eb_ia_i^3} + \frac{\alpha_s L}{Gb_ia_i} & 0 & 0 & 0 & \frac{6L^2}{Eb_ia_i^3} \\ 0 & 0 & \frac{4L^3}{Eb_ia_i^3} + \frac{\alpha_s L}{Gb_ia_i} & 0 & -\frac{6L^2}{Eb_i^3a_i} & 0 \\ 0 & 0 & 0 & \frac{\Delta\theta_x}{\Delta M_x} & 0 & 0 \\ 0 & 0 & -\frac{6L^2}{Eb_i^3a_i} & 0 & \frac{12L}{Eb_i^3a_i} & 0 \\ 0 & -\frac{6L^2}{Eb_i^3a_i} & 0 & 0 & 0 & \frac{12L}{Eb_i^3a_i} \end{pmatrix} \tag{3.50}$$

where L is the length of the micro-Euler–Bernoulli beam. a_i and b_i are the cross-sectional dimensions of the ith micro-Euler–Bernoulli beam and α_s is the shear coefficient of the material, which can be expressed as [6]:

$$\alpha_s = \frac{12 + 11\nu}{10(1+\nu)} \tag{3.51}$$

Table 3.1 Relationship between structural correction factors and structural parameters.

b/c	**1**	**2**	**10**	**∞**
k	0.141	0.229	0.312	0.333

The torsional model of the flexure hinge is developed taking into consideration the thickness-to-width ratio ($\tau = a_i/b_i$) and its equation can be expressed as [7]:

$$C_{\theta_x M_x} = \frac{7L}{2G}\left(\frac{1}{a_i b_i^3} + \frac{1}{a_i^3 b_i}\right)\frac{\tau^2 + 2.609\tau + 1}{1.17\tau^2 + 2.191\tau + 1.17} \tag{3.52}$$

where k is the structural correction factor of the flexure hinges, which is defined by b/c, and relationships are shown in Table 3.1.

In general, the proposed vibration stage is composed of many flexure hinges connected in parallel and series. Additionally, each flexure hinge can be considered as a series connection of all the microbeams, if the effects of stress distributions on elastic deformations of the flexure hinges are neglected. In order to develop the compliance of the vibration stage, the compliance of flexure hinges need to be transferred from the local coordinates o_r-$x_r y_r z_r$ to the global coordinates o_e–$x_e y_e z_e$ (see Figure 3.8):

$$C_r^e = \sum_{i=1}^{N} T_r^e C_i^r (T_r^e)^T \tag{3.53}$$

where C_r^e is the compliance at o_e–$x_e y_e z_e$ and T_r^e is the compliance transformation matrix (CTM), which can be expressed as [8–10]:

$$T_r^e = \begin{bmatrix} R_r(\delta) & S_r(r_r)R_r(\delta) \\ O & R_r(\delta) \end{bmatrix} \tag{3.54}$$

where $R_r(\delta)$ is the rotation matrix of the local coordinate o_r-$x_r y_r z_r$ transferring to the global coordinate o_e–$x_e y_e z_e$, which is given by:

$$R_r(\delta) = \begin{bmatrix} \cos\delta & \sin\delta & 0 \\ -\sin\delta & \cos\delta & 0 \\ 0 & 0 & 0 \end{bmatrix} \tag{3.55}$$

Here, $r_r = [x_r, y_r, z_r]$ is the position vector of the point o_r in the global coordinate o-xyz and $S_r(r_r)$ is the skew-symmetric operator for the position vector r_r, which can be defined as:

$$S_r = \begin{bmatrix} 0 - z_r & y_r & \\ z_r & 0 - x_r & \\ -y_r & x_r & 0 \end{bmatrix} \tag{3.56}$$

3.3.3 Compliance Modeling of Flexure Mechanism

Figure 3.9 shows a typical flexure mechanism including both series and parallel structures. The flexure mechanism end executor O - xyz is supported by several flexure chains (Ch_i)

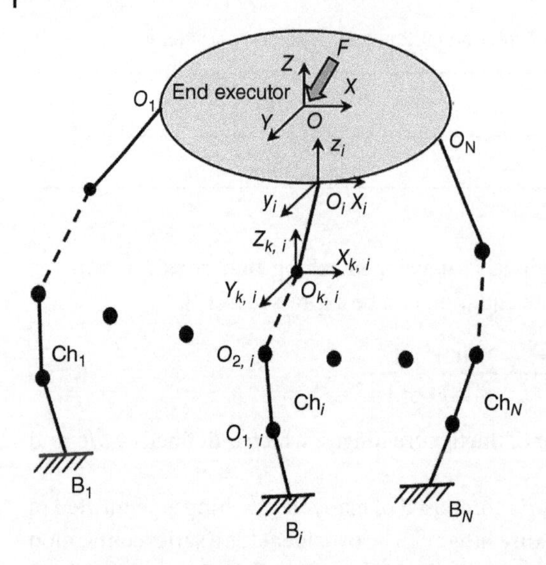

Figure 3.9 Working principles for the typical flexure mechanism.

in parallel, and each flexure chain consists of several flexure hinges connected in series. O_i - $x_iy_iz_i$ is the Cartesian coordinate system of the output point O_i for the ith flexure chain (Ch_i) and $O_{k,i}$ - $x^L_{k,i}y^L_{k,i}z^L_{k,i}$ is the Cartesian coordinate system for the kth flexure hinge on the ith flexure chain. If it is assumed that the compliance of the flexure hinge $O_{k,i}$ in the local coordinate system $O_{k,i}$ - $x^L_{k,i}y^L_{k,i}z^L_{k,i}$ is $C_{k,i}$, as it is transferred to the coordinate system O_i - $x_iy_iz_i$:

$$C^{o_i}_{k,i} = T^{o_i}_{k,i}C_{k,i}(T^{o_i}_{k,i})^T \tag{3.57}$$

where $C^{o_i}_{k,i}$ is the compliance for flexure hinge $O_{k,i}$ in the coordinate system $O_{k,}$ - $x^L_{k,i}y^L_{k,i}z^L_{k,i}$, $T^{o_i}_{k,i}$ is the compliance transformation matrix, which can be expressed as:

$$T^{o_i}_{k,i} = \begin{bmatrix} R^{o_i}_{k,i}(\alpha) & S^{o_i}_{k,i}(r_i)R^{o_i}_{k,i}(\alpha) \\ O & R^{o_i}_{k,i}(\alpha) \end{bmatrix} \tag{3.58}$$

where $R^{o_i}_{k,i}(\alpha)$ and $S^{o_i}_{k,i}(r_i)$ are the rotation matrix and position transformation matrix under the position vector $r_i = [x_i, y_i, z_i]$ for the flexure hinge $O_{k,i}$ to the flexure chain (Ch_i) coordinate O_i - $x_iy_iz_i$, respectively. These can be expressed as:

$$R_r(\alpha) = \begin{bmatrix} \cos\alpha & \sin\alpha & 0 \\ -\sin\alpha & \cos\alpha & 0 \\ 0 & 0 & 0 \end{bmatrix} \tag{3.59}$$

$$S^{o_i}_{k,i} = \begin{bmatrix} 0 - z_i & y_i & \\ z_i & 0 - x_i & \\ -y_i & x_i & 0 \end{bmatrix} \tag{3.60}$$

For the flexure chain (Ch_i) with k flexure hinges connected in series, the compliance C_{o_i} for the output point O_i relative to the fixed end (B_i) is:

$$C_{o_i} = \sum_{k=1}^{k} T^{o_i}_{k,i}C_{k,i}(T^{o_i}_{k,i})^T \tag{3.61}$$

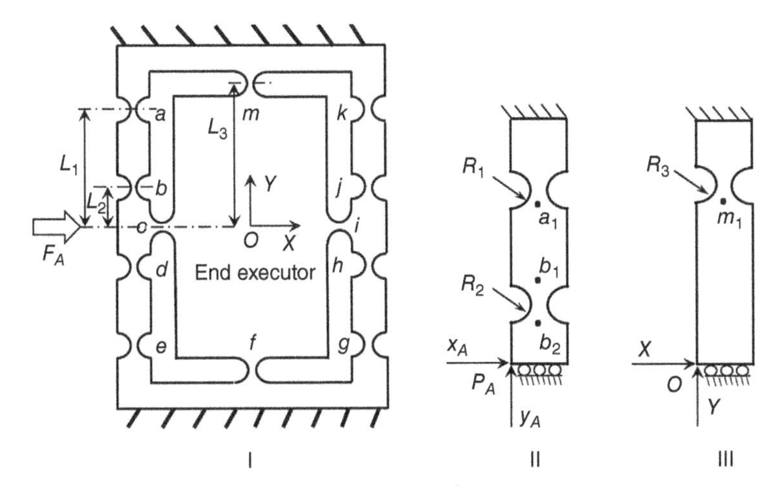

Figure 3.10 Compliances model of vibration stage.

Therefore, the compliance for the end executor C_o can be expressed as:

$$C_o = \left[\sum_{i=1}^{N}(T^o_{o_i}C_{o_i}(T^o_{o_i})^T)^{-1}\right]^{-1} \tag{3.62}$$

3.3.4 Compliance Modeling of the 2 DOF Vibration Stage

Because of the structure's symmetrical design, the output compliance C_{out}, which is defined as the compliance of the point o when an external force F_A is applied, is the same in both directions of the vibration stage. Moreover, an examination of the accuracy of the output compliance is always required to guarantee the accuracy. Figure 3.10I shows the simplified vibration stage's structural features and the main dimensional parameters. It consists of double-layer structures, with the outer layer constructed by the parallel connection of four bars and the inner layer constructed from normal right circle flexure hinges. Figure 3.10II shows one-quarter of the parallel connection of the four-bar structure for the outer layer structure including two right circle flexure hinges. Based on the matrix method, its output compliance C_{P_A} with respect to the global coordinates can be expressed as:

$$C_{P_A} = (k_{P_A})^{-1} = C^o_{a_1} + C^o_{b_1} + C^o_{b_2} = T^o_{P_A}C^{P_A}_{a_1}(T^o_{P_A})^T + T^o_{P_A}C^{P_A}_{b_1}(T^o_{P_A})^T + T^o_{P_A}C^{P_A}_{b_2}(T^o_{P_A})^T \tag{3.63}$$

where the local compliances $C^{P_A}_{a_1}$, $C^{P_A}_{b_1}$, and $C^{P_A}_{b_2}$ are the compliance at p_A-x_Ay_A and the corresponding global compliances $C^o_{a_1}$, $C^o_{b_1}$, and $C^o_{b_2}$, $T^o_{P_A}$ are the transformation matrices.

In order to further reduce the coupling motion, the compression/tensile stiffness of the flexure hinges in the inner layer is set much higher than their rotational stiffness. Therefore, the compression/tensile stiffness belonging to flexure hinges in the level direction can be ignored when calculating the inner structure's compliance. Because of the symmetrical design, the half inner layer structure only contains one flexure hinge as shown in Figure 3.3III and the relevant compliance C_o^{in} is:

$$C_o^{in} = (k_o^{in})^{-1} = C_{m_1}^o \tag{3.64}$$

where the global compliances $C_{m_1}^o$ do not need to be transferred.

Overall, the vibration stage contains three individual limbs connected at the center point o in parallel, including the structure of both the outer and inner layers. Therefore, the stage output compliance C_{out} can be calculated by considering the total compliance of the three limbs:

$$C_{out} = (k_{out})^{-1} = [2(2C_{P_A})^{-1} + (2C_{m_1}^o)^{-1}]^{-1} \tag{3.65}$$

Because of the symmetrical design and the locations of the placement of the piezoelectric actuators, the actuation forces are applied to the end executor directly. As the end executor is assumed to be rigid, the input and output points and relevant compliances are coincident and can be expressed as:

$$C_{in} = C_{out} \tag{3.66}$$

3.3.5 Dynamic Analysis of the Vibration Stage

As shown in Section 3.2.3, the structure in each degree of freedom (DOF) of the vibration stage is the same. Therefore, only unidirectional dynamics are discussed, and the dynamic model of the vibration stage is shown in Figure 3.11. Assuming that the end executor obtains a displacement of $s = [x, 0, 0]^T$ under the action force F, and the corresponding kinetic energy T_k and potential energy U_k can be expressed as:

$$T_k = \frac{1}{2}\dot{x}^2 M_2 + 2\dot{x}^2 M_1 + 2I_z\theta_z^2 \tag{3.67}$$

$$U_k = \frac{x^2}{2C_{out}} \tag{3.68}$$

where M_1 and M_2 denote the equivalent mass of flexure hinges and the equivalent mass of the end executor, respectively, and I_z and $\theta_z = \frac{x}{l}$ are the moment of inertia of the flexure hinges around the z-axis and the corresponding turning angles, respectively.

The dynamics of the vibration stage can be obtained by substituting the kinetic and potential energies into the Lagrangian equation:

$$\frac{d}{dt}\left(\frac{\partial T_k}{\partial \dot{s}}\right) - \frac{\partial T_k}{\partial \dot{s}} + \frac{\partial U_k}{\partial s} = F \tag{3.69}$$

Overall, the dynamic equation for describing the free motion of the vibration stage can be expressed as:

$$\left(M_2 + 4M_1 + \frac{4I_z}{l^2}\right)\ddot{x} + \frac{x}{C_{out}} = F \tag{3.70}$$

Then, the natural frequency of the vibration stage can be obtained as:

$$\omega = \sqrt{\frac{1}{\left(M_2 + 4M_1 + \frac{4I_z}{l^2}\right)C_{out}}} \tag{3.71}$$

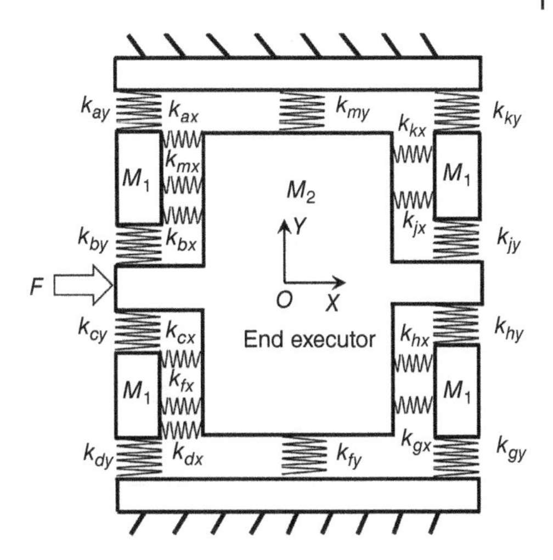

Figure 3.11 Dynamic model of the vibration stage.

3.3.6 Finite Element Analysis of the Mechanism

3.3.6.1 Structural Optimization

In order to achieve the best performance of the developed vibration stage under the condition of being mutually restrictive, the structure dimensional parameters of the vibration stage are optimized by importing the analytical model to the ANSYS mechanical APDL software. This aims to achieve both as large as possible a natural frequency and the greatest level of compliance. According to the overall structural design, the range of R_1, R_2, and R_3 is shown below:

$$\begin{cases} 0.5 < R_1 < 2.9 \\ 0.5 < R_2 < 2.9 \\ 2 < R_3 < 5 \end{cases} \tag{3.72}$$

In addition, the mathematical model can be expressed as:

$$\begin{cases} f_{max} = f(R_1, R_2, R_3) \\ C_{max} = C(R_1, R_2, R_3) \end{cases} \tag{3.73}$$

As a result, the optimized structural parameters and the material properties are listed in Table 3.2.

3.3.6.2 Static and Dynamic Performance Analysis

In order to evaluate the new analytical model, the output compliance and natural frequency of the vibration stage are validated using the finite element analysis (FEA) method via ANSYS mechanical APDL. The meshing of the vibration stage is based on the hexahedron-dominated method, while the approximate and curvature control method in the advanced dimension control is selected for the meshing of the curve structure.

The static analysis is first performed, where a 1 N constant force is applied to the vibration stage and the results are shown in Figure 3.12. It can be seen in Figure 3.12a that a maximum deformation of 0.02 μm occurs at the flexure hinges. This corresponds to the structural stiffness of the vibration stage of 50 N/μm or a structural compliance of 0.02 μm/N. As shown in Table 3.2, the discrepancy between the analytical model and the FEA model is 3.5%, which

Table 3.2 Vibration stage structural parameters and material properties used in the FEA model.

Structural parameters								
R_1 (mm)	R_2 (mm)	R_3 (mm)	L_1 (mm)	L_2 (mm)	L_3 (mm)	a (mm)	b (mm)	c (mm)
1.71	1.14	4.96	34	9.5	31	9	6	2

Material properties			
Density	Young's modulus	Poisson's ratio	Yield strength
7850 kg/m^3	197.5 GPa	0.3	430 MPa

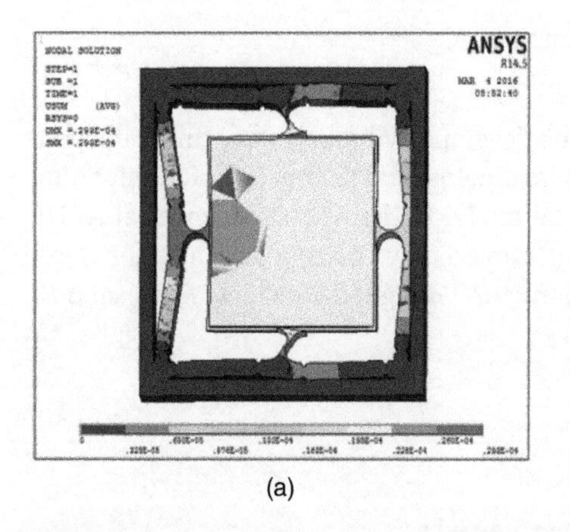

(a)

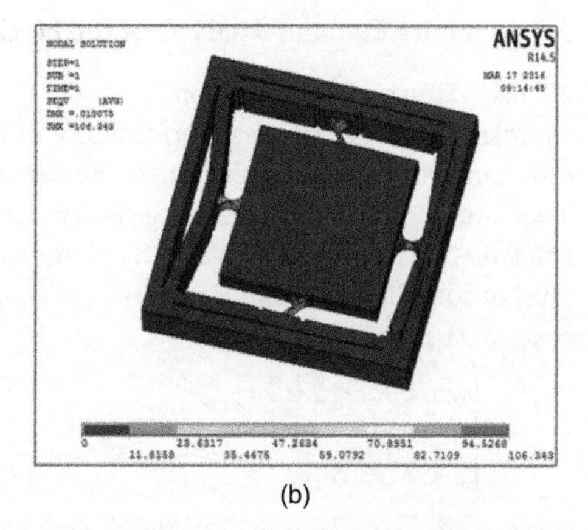

(b)

Figure 3.12 Static FEA simulation of the vibration stage: (a) displacement distribution and (b) stress distribution.

indicates the high accuracy of the analytical model. The error may be attributed to the complex shear and stress concentration effects in the connections of the flexure hinges, and this can be compensated by closed loop control of the vibration stage. Figure 3.12b shows the stress distribution of the vibration stage, and it can be noticed that the maximum stress value is 106.3 MPa, which is well below the yield strength of the material (430 MPa).

Finite element (FE) modal analysis is performed to evaluate the dynamic performance of the developed vibration stage. Figure 3.13 shows the shapes of the first two vibrational modes, and the corresponding natural frequencies are 2760.4 and 10 661.7 Hz, respectively. In the first vibration mode shape, it can be seen that the vibration stage moves up and down along the z-axis (outside of the plane). For the second vibration mode shape, the vibration stage moves back and front along the y-axis. The deviations shown in Table 3.3 between the analytical and FEA modal for the first and second natural frequencies are 1.78% and 2.2%, respectively. This shows that the developed analytical model can accurately predict the dynamic performance of the vibration stage.

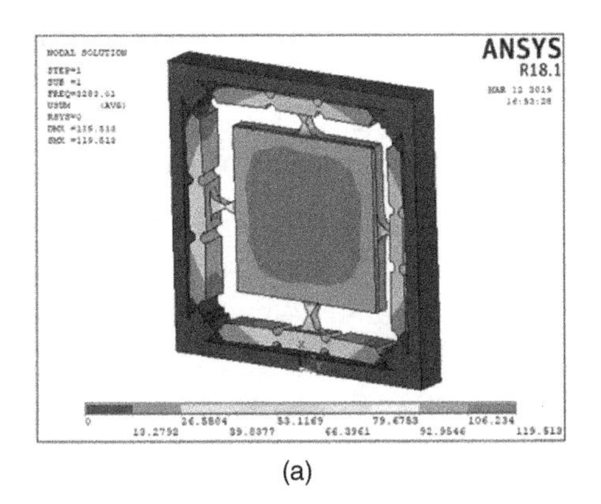

(a)

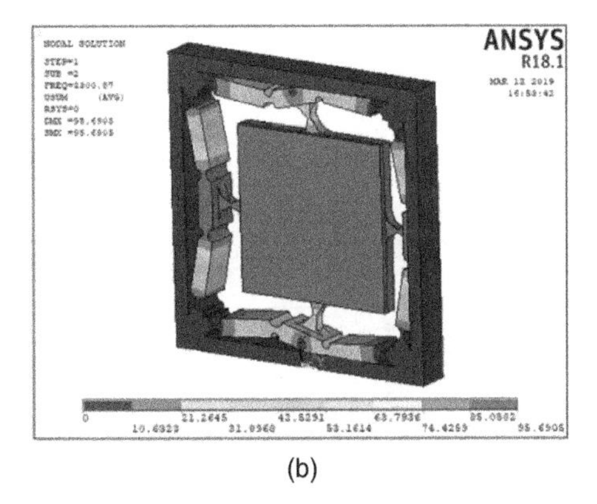

(b)

Figure 3.13 The first two modes of the vibration stage: (a) the first mode of deformation along the *z*-axis (2760.4 Hz) and (b) the second mode of deformation along the *y*-axis (10661.7 Hz).

Table 3.3 Comparisons of static and dynamic results.

	Compliance	First resonance frequency	Second resonance frequency
FEA	0.02 μm/N	2760.4 Hz	10 661.7 Hz
Analytical model	0.0193 μm/N	2809.5 Hz	10 896.3 Hz
Deviation (%)	3.5	1.78	2.2

Table 3.4 Specifications of P-844.20 piezo actuators.

Travel range	30 μm	Electrical capacitance	12 μF
Resolution	0.3 nm	Resonant frequency	12 kHz
Stiffness	107 N/μm	Mass without cable	108 g
Push/pull force	3000 N	Length	65 mm

3.3.7 Piezoelectric Actuator Selection

Two preloaded piezoelectric actuators (P-844.20 Physik Instrumente) are selected to drive the proposed vibration stage, taking into account the working frequency and amplitude range. The specifications are given in Table 3.4.

The actually achievable displacement of the vibration stage can be expressed as:

$$\Delta L = \Delta L_0 \frac{K_a}{K_a + K_{out}} \tag{3.74}$$

where ΔL and ΔL_0 are the displacements of the end executor and the stroke of the piezoelectric actuators, respectively. K_a is the stiffness of the piezoelectric actuators, which is 107 N/μm, and K_{out} represents the stiffness of the vibration stage. Therefore, the maximum displacement of the vibration stage is almost 20 μm, which is big enough for further application.

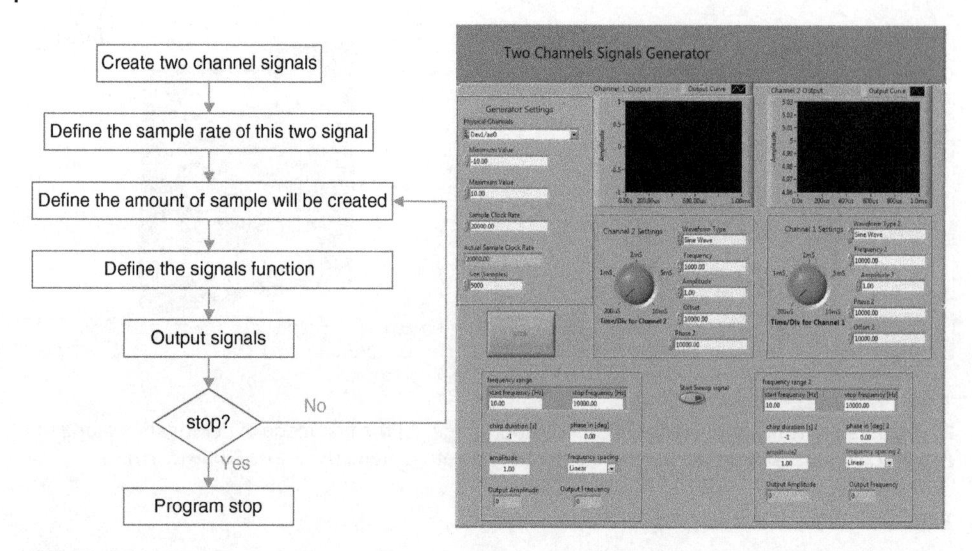

Figure 3.14 The architecture and LabVIEW control panel of the control signal generation part. Source: LabVIEW control panel.

3.3.8 Control System Design

3.3.8.1 Control Program Construction

The control system is developed based on the LabVIEW platform (2014 SP1) because of its friendly graphical user interface, powerful data acquisition and analysis capabilities, and highly integrated control of system-related equipment. As the vibration stage is driven by two separate piezoelectric actuators, two individual continuous control signals are needed, and data for real-time stage vibration displacements also need to be collected and recorded. Therefore, the developed control system consists of two parts involving control signal generation by a signal generator and data collection by an oscilloscope. Figure 3.14 shows the architecture of the control signal generation. The output signals and their sample rate, which is mainly affected by the output hardware, need to be defined first. Because LabVIEW software does not support an infinite generation sample function, then a certain number of samples should be defined. In order to obtain continuous output signals, a structure named "for loop" is applied here, and the program will cycle the function infinitely until "stop" button is pressed. Other functions such as generating a sweep signal are also added to the program. Figure 3.15 shows the architecture of the data collection program part. Similarly, the data collection channels and sample rate need to be set first. The "for loop" function is utilized to collect the data infinitely. After a series of mathematical operations and filtering processes, vibration displacement data can be obtained.

3.3.9 Hardware Selection

In the corresponding hardware system, the data acquisition (DAQ) device acts as the interface between the program and signals from the outside world. Two types of DAQ devices are widely used, which are a data output module to generate signals and a data input module to digitize the incoming signals so that a computer can interpret them. Because the

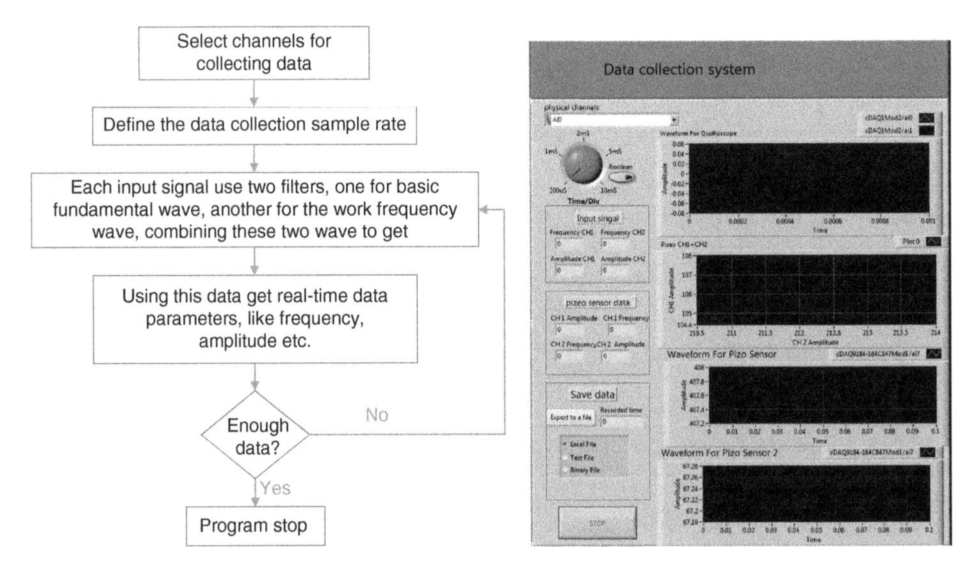

Figure 3.15 The architecture and LabVIEW control panel of the data collection part. Source: LabVIEW control panel.

Table 3.5 Specifications of selected DAQ devices.

Production name	Module type	Signal ranges	Channels	Sample rate	Resolution	Isolation	Connectively
NI 9221	Voltage input	±60 V	8 Single-ended	800 kS/s	12-Bit	250 Vrms Ch-Earth (Screw Terminal) 60 VCD Ch-Earth (D-SUB)	Screw terminal 25-Pin D-SUB
NI 9263	Voltage output	±10 V	4	100 kS/s/ch	16-Bit	250 Vrms Ch-Earth	Screw terminal

proposed vibration system aims to work under thousands of vibration frequencies, DAQ devices with low sample rates will not be able to truly restore the input or output signals and could not generate or receive the desired vibration amplitude and frequency. Taking into consideration the sample rate and resolution issues, the DAQ card selection results are shown in Table 3.5. Moreover, the two cards were mounted onto the cDAQ-9178, which is an Eight-Slot USB chassis. It is designed for small, portable sensor measurement systems. The chassis provides the plug-and-play simplicity of USB for sensor and electrical measurements, and the data transfer between the cards and the computer can be optimized.

To effectively monitor the vibration displacement, a modular capacitive sensor system (DL6220, Micro-epsilon) with two capacitive sensors (CS005, Micro-epsilon) is selected and their technical parameters are shown in Table 3.6.

Table 3.6 Specifications of selected capacitance sensors and controller.

Demodulator DL6200					
Resolution dynamic	**Bandwidth**	**Data rate digital output**	**Sensitivity deviation**	**Analog output**	**No. of channels**
0.02% FSO	5 kHz (−3 dB)	3.906 kSa/s	≤±0.1% FSO	0–10 V	Max. 4

Capacitance sensor CS005					
Measuring range	**Linearity**		**Resolution (nm)**		**Dimensions**
0.05 mm	≤±0.15 μm	≤±0.3% FSO	0.0375 (static)	1 (dynamic)	Ø6 × 12 mm

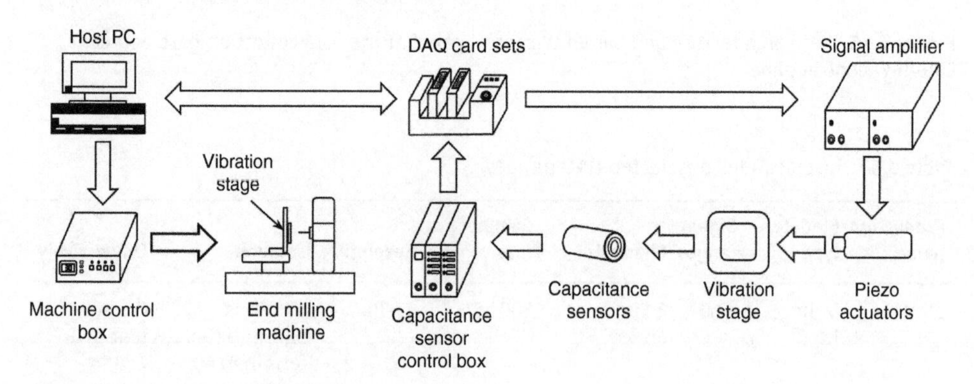

Figure 3.16 Layout of the control system.

3.3.10 Layout of the Control System

The schematic of the whole system is shown in Figure 3.16. Both the machine tool and vibration stage are controlled by the same host computer. For the vibration stage, the parameters of control signals are set in the LabVIEW program and sent from an NI 9263 output DAQ (data acquisition) card. They can be used to drive the piezo actuators by going through a high-voltage piezo amplifier. At the same time, the vibration displacement is monitored by a modular capacitive sensor system (DL6220, Micro-epsilon) with two capacitive sensors (CS005, Micro-epsilon) in each direction. A NI 9221 input DAQ card is selected for collecting the stage displacement data from the capacitive sensors.

3.4 Concluding Remarks

This chapter presents the theory and principles of vibration system design, which mainly include the resonant and nonresonant vibration systems. The design process of the resonant vibration system mainly includes the power calculation, transducer, and horn design. The relationship between cutting energy and vibration system power is discussed and the vibration mode and parameters of piezoelectric ceramics are expounded to provide the necessary basis for the selection of piezoelectric ceramics. The wave equation of one-dimensional longitudinal vibration of variable cross-sectional bar is proposed, and the theoretical calculation model of the structural parameters of the vibration transducer and horn is obtained based on this. A half-wavelength ultrasonic vibrator is proposed, its transducer adopts a sandwich structure, and the quarter-wavelength composite structure is chosen for the horn.

In the nonresonant vibration system part, the examples of the design and analysis of a vibration device are presented. To achieve high motion accuracy and to reduce the coupling effect between the two vibration directions, the vibration stage is designed to be completely symmetrical based on double-parallel four-bar linkages with a double-layer flexure hinge structure. The compliance of the flexure mechanisms is modeled using the matrix-based compliance modeling method. The dynamic characteristics and complete compliance are carried out by using the Lagrangian principle and considering the stage structure, respectively. Acceptable deviations (3.5% for compliance, 1.78% for first resonance frequency and 2.2% for second resonance frequency) can be obtained by comparing the analytical results with the FEA results, which indicate the feasibility of the proposed analytical model.

References

1 Li, Y. and Xu, Q. (2009). Design and analysis of a totally decoupled flexure-based XY parallel micromanipulator. *IEEE Trans. Rob.* 25: 645–657. https://doi.org/10.1109/tro.2009.2014130.

2 Zhu, Z.W., Zhou, X.Q., Wang, R.Q., and Liu, Q. (2014). A simple compliance modeling method for flexure hinges. *Sci. China Technol. Sci.* 58: 56–63. https://doi.org/10.1007/s11431-014-5667-1.

3 Li, Y. and Xu, Q. (2009). Design and analysis of a totally decoupled flexure-based XY parallel micromanipulator. *IEEE Trans. Rob.* https://doi.org/10.1109/TRO.2009.2014130.

4 Ryu, J.W., Lee, S.Q., Gweon, D.G., and Moon, K.S. (1999). Inverse kinematic modeling of a coupled flexure hinge mechanism. *Mechatronics* 9: 657–674. https://doi.org/10.1016/S0957-4158(99)00006-9.

5 Lobontiu, N., Paine, J.S.N., Garcia, E., and Goldfarb, M. (2001). Corner-filleted flexure hinges. *J. Mech. Des.* https://doi.org/10.1115/1.1372190.

6 Cowper, G.R. (1966). The shear coefficient in Timoshenko's beam theory. *J. Appl. Mech.* 33: 335. https://doi.org/10.1115/1.3625046.

7 Chen, G. and Howell, L.L. (2009). Two general solutions of torsional compliance for variable rectangular cross-section hinges in compliant mechanisms. *Precis. Eng.* 33: 268–274. https://doi.org/10.1016/j.precisioneng.2008.08.001.

8 Tang, H. and Li, Y. (2013). Design, analysis, and test of a novel 2-DOF nanopositioning system driven by dual mode. *IEEE Trans. Rob.* 29: 650–662. https://doi.org/10.1109/TRO.2013.2248536.

9 Koseki, Y., Tanikawa, T., Koyachi, N., and Arai, T. (2002). Kinematic analysis of a translational 3-d.o.f. micro-parallel mechanism using the matrix method. *Adv. Rob.* 16: 251–264. https://doi.org/10.1163/156855302760121927.

10 Pham, H.H. and Chen, I.M. (2005). Stiffness modeling of flexure parallel mechanism. *Precis. Eng.* 29: 467–478. https://doi.org/10.1016/j.precisioneng.2004.12.006.

11 Zheng, L., Chen, W., Pozzi, M. et al. (2018). Modulation of surface wettability by vibration assisted milling. *Precis. Eng.*: 1–10. https://doi.org/10.1016/j.precisioneng.2018.09.006.

12 Zheng, L., Chen, W., and Huo, D. (2018). Experimental investigation on burr formation in vibration-assisted micro-milling of Ti–6Al-4V. *Proc. Inst. Mech. Eng. Part C J. Mech. Eng. Sci.* https://doi.org/10.1177/0954406218792360.

13 Chen, W., Zheng, L., and Huo, D. (2018). Surface texture formation by non-resonant vibration assisted micro milling. *J. Micromech. Microeng.* 28: 025006.

3.A Appendix

```
**General Preprocessing**
/PREP7

ET, 1, SOLID98,3              ! 3-D COUPLED-FIELD SOLID !
ET, 2, SOLID95,               !DEFINE ELEMENT TYPE FOR THE BACK
COVER BOARD
OR POLE
**matrial properties for steel**
MP, DENS, 1, 7800                        ! STEEL DENSITY, KG/M**3
MP, EX, 1, 2.1E11                         ! ELASTIC MODULI, N/M^2
MP, PRXY, 1, 0.3                             ! POISSON'S RATIO
**matrial properties for aluminum**
!MP, DENS, 2, 2700
! ALUMINUM DENSITY
!MP, EX, 2, 7.1E10
! ALUMINUM ELASTIC MODULI
!MP, PRXY, 2, 0.345
! ALUMINUM POISSON'S RATIO

**matrial properties for piezoceramic1**
mp,dens,2,7500  !desity, kg/m^3
mp,perx,2,6.74565e-9  !permittivity at constant strain, F/m
mp,pery,2,6.74565e-9
mp,perz,2,5.86687e-9
tb,anel,2,6,,0,   !anisotropic elastic stiffness, N/m^2
tbdata,1,1.39e11,7.78e10,7.43e10  !c11,c12,c13
tbdata,7,1.39e11,7.43e10          !c11,c13
tbdata,12,1.15e11                 !c33
tbdata,16,3.06e10                 !c44
tbdata,19,2.56e10                 !c55
tbdata,21,2.56e10                 !c66
```

```
tb,piez,2   !piezoceramic stress coefficients, C/m^2
tbdata,3,-5.20279     !e13
tbdata,6,-5.20279     !e23
tbdata,9,15.08041     !e33
tbdata,14,12.71795     !e42
tbdata,16,12.71795     !e51
```

```
   CYLIND, 0.025, 0.01, 0, 0.005, 0, 360
! THE BACK COVER BOARD
   WPOFFS, 0, 0, 0.005
   CYLIND, 0.025, 0.01, 0, 0.005, 0, 360
! PIEZOELECTRIC CERAMIC
   WPOFFS, 0, 0, 0.005
   CYLIND, 0.025, 0.01, 0, 0.005, 0, 360
! PIEZOELECTRIC CERAMIC
   WPOFFS, 0, 0, 0.005
   CYLIND, 0.025, 0.01, 0, 0.005, 0, 360
! PIEZOELECTRIC CERAMIC
   WPOFFS, 0, 0, 0.005
   CYLIND, 0.025, 0.01, 0, 0.005, 0, 360
! PIEZOELECTRIC CERAMIC
   WPOFFS, 0, 0, 0.005
   CYLIND, 0.035, 0.01,0, 0.005, 0, 360
! THE FROND COVER BOARD
   WPOFFS, 0, 0, 0.005
   CONE,0.025,0.01,0,0.015,0,360,
   WPOFFS, 0, 0, 0.015
   !CYLIND, 0.025, 0,0, 0.003, 0, 360
! THE FROND COVER BOARD
   !WPOFFS, 0, 0, 0.003
   CYLIND, 0.01, 0,0, 0.03, 0, 360
   VGLUE,ALL
   NUMCMP, ALL

    !DESTINATION ENTITY ATTRIBUTE
   VSEL,S, , ,        1
   VATT, 1 , , 2, ,
   VSEL,S, , ,        4
   VATT, 2 , , 1, ,
   VSEL,S, , ,        5
   VATT, 2 , , 1, ,
```

```
VSEL,S, , ,        6
VATT, 2 , , 1, ,
VSEL,S, , ,        7
VATT, 2 , , 1, ,
VSEL,S, , ,        2
VATT, 1 , , 2, ,
VSEL,S, , ,        3
VATT, 1 , , 2, ,
VSEL,S, , ,        8
VATT, 1 , , 2, ,
VSEL,ALL

SMRTSIZE, 6
MSHAPE, 1,3D
MSHKEY, 0
VMESH, ALL
NUMMRG, ALL, , , , LOW
NUMCMP, ALL
CSYS, 0
WPCSYS, 0
ALLSEL, ALL, ALL
FINI
/SOL
!*
ANTYPE,2
!*
!*
MODOPT,LANB,20
EQSLV,SPAR
MXPAND,20, , ,0
LUMPM,0
PSTRES,0
!*
MODOPT,LANB,20,200,40000, ,OFF
FLST,2,2,5,ORDE,2
FITEM,2,11
FITEM,2,-12
!*
/GO
DA,P51X,ALL,

SOLVE
FINI
```

4

Kinematics Analysis of Vibration-Assisted Machining

4.1 Introduction

Tool–workpiece separation (TWS) is recognized as one of the main reasons for the benefits of vibration-assisted machining (VAM). To accelerate the progress of the application of VAM in industrial production, it is essential to systematically study the kinematics of VAM. On this basis, the underlying mechanism of each type of TWS involved in VAM and the required conditions could be determined. This would contribute to the establishment of the technological principles for the optimization of the machining parameters in VAM, thereby achieving the targeted machining quality.

The generation of TWS is directly related to the specific type of machining process used in VAM. As introduced in Chapter 1, different machining processes could be employed to enable the VAM. In turning, it is usually straightforward to determine the TWS because of the simple trajectory of the tool. However, because of the continually changed cutting velocity and uncut chip thickness (UCT) in milling, the mechanism of TWS in vibration-assisted milling (VAMILL) tends to be more complex. Previous research has experimentally demonstrated the feasibility and benefits of VAMILL in improving the machinability of hard materials, but the kinematics, TWS, and cutting mechanism of VAMILL are still not well understood.

As for vibration-assisted grinding, many abrasive grains would be synchronously involved in cutting, and so the trajectories of the abrasives become even more complex. Usually, one or two adjacent abrasive particles are selected to investigate the vibration grinding process, which is similar to milling.

Overall, there is a significant knowledge gap on the mechanism of TWS generation in VAM, which hinders the application and transfer of VAM from academia to industry. For these reasons, this chapter intends to fill this gap via a systematic approach to the kinematics and typical TWS mechanisms in VAM. The outcome will offer technological guidance for the determination of the optimal machining and vibration parameters and vibration-assisted system design.

Vibration Assisted Machining: Theory, Modelling and Applications,
First Edition. Lu Zheng, Wanqun Chen, and Dehong Huo.

Table 4.1 Turning cutting types.

Cutting types	Vibration of tool tip	Cutting types	Tool tip displacement
Conventional cutting			$\begin{cases} x(t) = vt \\ z(t) = 0 \end{cases}$
CDVA cutting	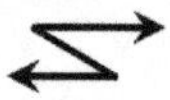	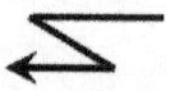	$\begin{cases} x(t) = vt + a \cos(2\pi ft) \\ z(t) = 0 \end{cases}$
NDVA cutting		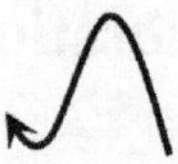	$\begin{cases} x(t) = vt \\ z(t) = b \sin(2\pi ft) \end{cases}$
EVA cutting			$\begin{cases} x(t) = vt + a \cos(2\pi ft) \\ z(t) = b \sin(2\pi ft) \end{cases}$

4.2 Kinematics of Vibration-Assisted Turning

As listed in Table 4.1, the vibration-assisted turning process, depending on the vibration trajectory characteristics of the tool tip, can be divided into three types, including [1]

(i) Cutting-directional vibration-assisted (CDVA) cutting, in which the tool vibrates in the cutting direction only;
(ii) Normal-directional vibration-assisted (NDVA) cutting, in which the tool vibrates only in the direction normal to the cutting direction;
(iii) Elliptic vibration-assisted (EVA) cutting, in which the tool vibrates simultaneously in both the cutting and normal directions.

Essentially, EVA cutting is a 2D cutting process. CDVA and NDVA cuttings are 1D VAM cutting, and they can be regarded as the special cases of EVA cutting.

Figure 4.1 schematically illustrates the motion trajectories of a tool tip in a vibration-assisted turning process. Here, the principal cutting direction is set in the positive x-direction, while the depth of cut is normal to the uncut surface, namely, the z-direction. The y-direction in Figure 4.1 is along the cross-feed direction (equivalent to the feed direction in a turning operation). The tool's primary feed rate is represented by v and the depth of cut is δ.

Here, the tool vibration trajectory could be described as follows:

$$\begin{cases} x_{Tool}(t) = a \cos(2\pi ft) \\ z_{Tool}(t) = b \sin(2\pi ft) \end{cases} \tag{4.1}$$

where a and b are the amplitudes of vibration in the x- and z-directions and f is the vibration frequency.

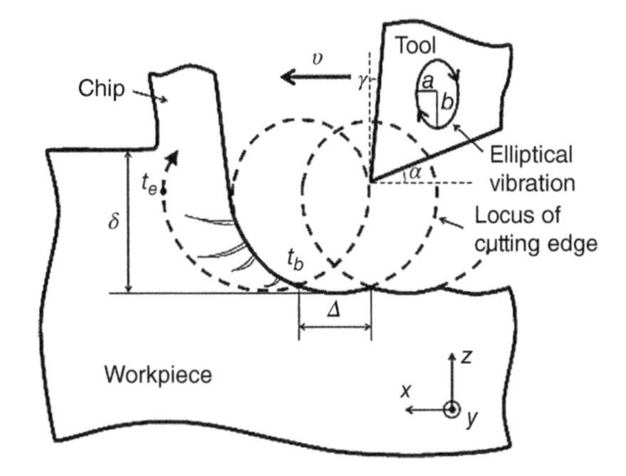

Figure 4.1 A schematic illustration of vibration-assisted cutting. Source: Xu and Zhang [1].

The motion of the tool tip relative to workpiece in vibration-assisted turning could be expressed as:

$$\begin{cases} x(t) = vt + a\cos(2\pi ft), \\ z(t) = b\sin(2\pi ft), \end{cases} \begin{pmatrix} \text{CDVA cutting}: a \neq 0, \quad b = 0 \\ \text{NDVA cutting}: \text{a} = 0, \quad b \neq 0 \\ \text{EVA cutting}: a \neq 0, \quad b = 0 \end{pmatrix} \tag{4.2}$$

Based on Eq. (4.2), the speed of the tool tip motion relative to the workpiece is

$$\begin{cases} v_x(t) = v - 2\pi fa\sin(2\pi ft), \\ v_z(t) = 2\pi fb\cos(2\pi ft), \end{cases} \begin{pmatrix} \text{CDVA cutting}: a \neq 0, \quad b = 0 \\ \text{NDVA cutting}: a = 0, \quad b \neq 0 \\ \text{EVA cutting}: a \neq 0, \quad b = 0 \end{pmatrix} \tag{4.3}$$

To clearly analyze the kinematics of vibration-assisted turning, in the following sections, the mechanism of TWS in 1D and 2D VAM turning is discussed.

4.2.1 TWS in 1D VAM Turning

In 1D VAM, as illustrated in Figure 4.2, the horizontal vibration would be imposed on the cutting tool and/or workpiece for a typical cutting system consisting of workpiece and cutting tool.

At time t_1, the cutting process has started, and the tool rake face has come into direct contact with the uncut materials, as shown in pane 1 of 4.2. Here, the tool velocity $v_x(t)$ relative to workpiece is regarded as greater than zero. With the advance of cutting tool, it

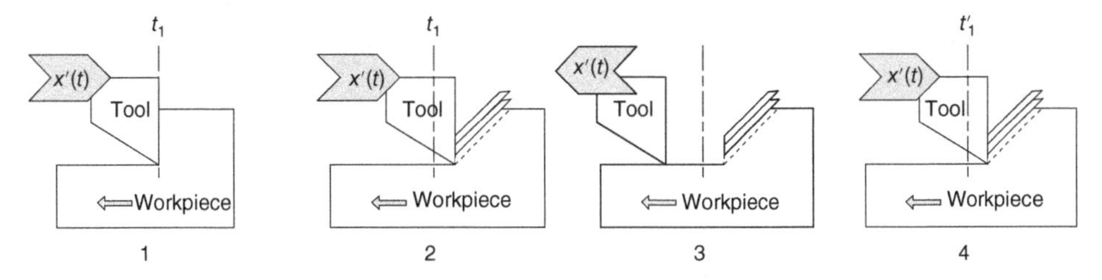

Figure 4.2 1D vibration-assisted machining.

has reached its maximum value of horizontal linear vibration path at time t_2, when the tool velocity $v_x(t)$ has decreased to zero and it is about to reverse its movement direction. In panel 3, $v_x(t)$ has reversed, and the tool withdraws from the workpiece. In panel 4, $v_x(t)$ becomes positive again, and the tool is shown to be advancing toward contact with the workpiece as it commences another cutting cycle, with t_1 representing the time at which the cutting tool initially makes contact with the uncut workpiece material at a new location. The duration of the full cycle is T, which is equal to $1/f$. The portion of the cycle in which the tool is cutting is $t_2 - t_1$.

As indicated in Figure 4.2, the intermittent contact between the cutting tool and chip can be defined by two-time variables: t_1 when the tool enters the uncut work material and t_2 when it separates.

$$A\ \sin(2\pi f t_1) = A\ \sin\left(\arccos\left(\frac{-V}{2\pi fA}\right)\right) - V\left(t_1 - \frac{\arccos(-V/2\pi fA)}{2\pi f}\right) \tag{4.4}$$

$$t_2 = \frac{\arccos(-V/2\pi fA)}{2\pi f} \tag{4.5}$$

Equation (4.4) must be solved numerically. The up-feed increment (F_{UP}) and horizontal speed ratio (HSR) are defined as:

$$F_{UP} = \frac{V}{f} \tag{4.6}$$

$$\text{HSR} = \frac{V}{2\pi fA} \tag{4.7}$$

where F_{UP} is the distance between the equivalent point on the tool vibration path for successive cycles. It is equal to the distance traveled by the tool in one vibration cycle, relative to the workpiece. HSR is the ratio between the workpiece up-feed velocity and the peak horizontal vibration speed of the tool. In general, non-interrupted cutting would occur when HSR $\geq$ 1.0. Both F_{UP} and HSR are useful parameters in characterizing a VAM machining cycle.

For 1D VAM, a "duty cycle," DC_1, can be defined based on the portion of each vibration cycle when the tool is cutting the workpiece [2, 3].

$$\text{DC}_1 = \frac{t_2 - t_1}{T} = f(t_2 - t_1) \tag{4.8}$$

where the period of the tool vibration cycle is T (equal to $1/f$) and the variables t_1, t_2, and f had been defined above. A larger value of the duty cycle indicates that the tool is cutting for a larger proportion of each cycle.

Figure 4.3 shows a duty cycle as a function of HSR. In conventional machining as well as non-interrupted 1D VAM, the tool rake face would continuously be in contact with the workpiece, and the duty cycle is 1.0.

In addition, there is a critical up-feed velocity V_{crit}, which can be calculated using the following equation:

$$V_{crit} = \omega A = 2\pi fA \tag{4.9}$$

When the up-feed velocity is larger than V_{crit}, the tool rake face would never separate from the uncut work surface.

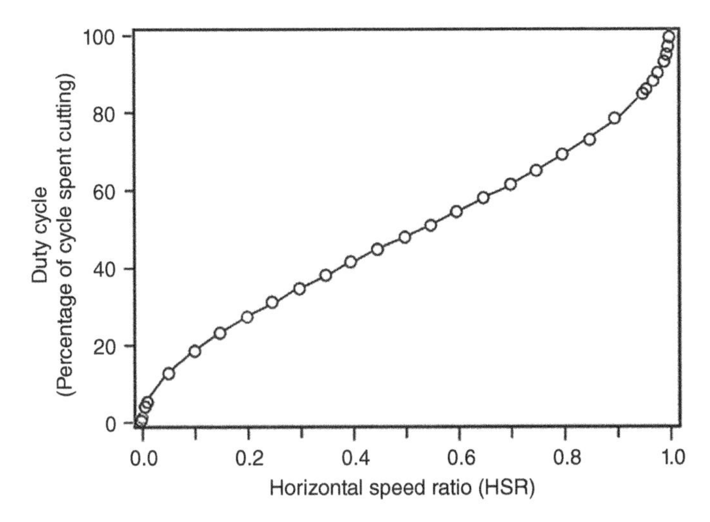

Figure 4.3 1D VAM duty cycle chart. Source: Cerniway [2].

Figure 4.4 Relative speed curve of ultrasonic honing.

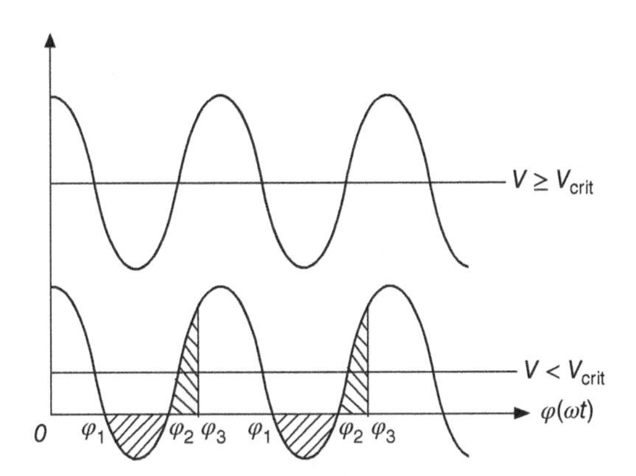

Figure 4.4 shows the relative speed curve of ultrasonic honing. If $V < V_{crit}$, periodic intermittent cutting would take place. For the interval from φ_1 to φ_2, there is a negative relative speed between the tool and workpiece. This means that the tool starts to separate the workpiece from the point φ_1 and the distance between the tool and the workpiece gradually increased. When φ reaches φ_2, the tool starts to approach the workpiece until φ reaches φ_3, when it is close to the workpiece again, and resumes the cutting process.

When $V \geq V_{crit}$, continuous cutting occurs, despite the harmonic variation of the relative velocity between the tool and workpiece. Most of the advantages of VAM, such as extended tool life and improved surface finish, are mainly ascribed to the periodic separation between the tool rake face and the uncut material. Thus, the relative speed V should be smaller than V_{crit}, in order to achieve interrupted cutting.

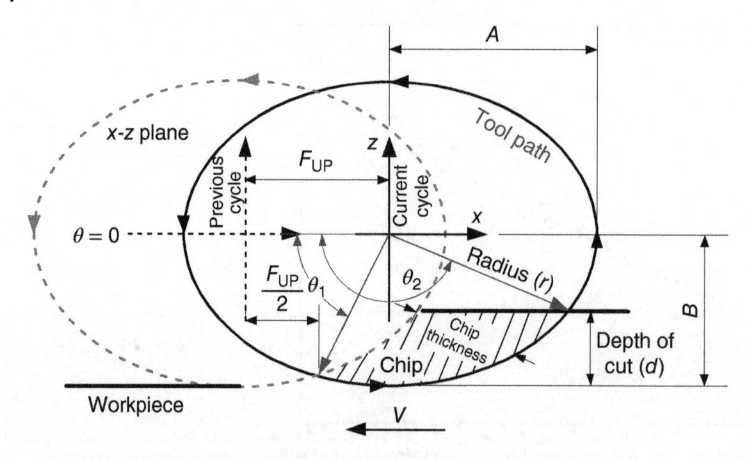

Figure 4.5 Schematic illustration of 2D VAM.

4.2.2 TWS in 2D VAM Turning

In 2D VAM, the vertical harmonic motion should be imposed onto the horizontal motion of 1D VAM, thereby enabling the tool tip to move in a tiny circle or ellipse. As illustrated in Figure 4.5, when the movement distance per cycle of workpiece is smaller than or comparable to the horizontal vibration amplitude, 2D VAM can be approximately depicted as a series of overlapping ellipses.

The relative position of the cutting tool to the workpiece in 2D VAM can be written as:

$$x(t) = -A\ \cos(\omega t) + Vt \tag{4.10}$$

$$z(t) = -B\ \sin(\omega t) \tag{4.11}$$

where the vertical position and velocity of the cutting tool relative to the workpiece are $z(t)$ and $z'(t)$, respectively, while B is the amplitude of vertical vibration. Other variables are as defined for 1D VAM. A and B are the lengths of the semimajor and semiminor axes of the tool path ellipse.

The tool velocity relative to the workpiece can be written as:

$$x'(t) = \omega A\ \sin(\omega t) + V \tag{4.12}$$

$$z'(t) = -\omega\ \cos(\omega t) \tag{4.13}$$

As shown in Figure 4.5, there is an angular position θ, with θ equal to ωt. Equations (4.9)–(4.12) are referenced to $\theta = 0$.

The instantaneous direction of the tool motion relative to the workpiece, $\aleph(t)$, is given by:

$$\aleph(t) = \arctan\left(\frac{-\omega B\cos(\omega t)}{\omega A\sin(\omega t) + V}\right) \tag{4.14}$$

and the instantaneous rake face angle $\gamma(t)$ and clearance angle $\alpha(t)$, relative to the workpiece, are

$$\gamma(t) = \gamma_0 + \aleph(t) \tag{4.15}$$

$$\alpha(t) = \alpha_0 - \aleph(t) \tag{4.16}$$

where γ_0 and α_0 are the tool rake and clearance angles of the cutting tool. VAM parameters should be optimized to ensure that the tool flank face does not touch the workpiece material during the downward portion of the elliptical tool path. When Eq. (4.16) is solved for $t = t_1$, $\alpha(t)$ should be equaling to or larger than zero. At time t_1, the tool contacts the workpiece during the downward portion of the vibration cycle (ωt_1 corresponds to θ_1 in Figure 4.5).

The critical up-feed velocity V_{crit}, up-feed increment (F_{UP}), and HSR are the same as given for 1D VAM in Eqs. (4.3), (4.6), and (4.7). In this chapter, only interrupted cutting in 2D VAM (HSR < 1) is reviewed.

The duty cycle in 2D VAM refers to the length of the elliptical tool path spent in contact with the workpiece material, compared to the perimeter of the ellipse [2], as defined in Eq. (4.17):

$$DC_2 = \frac{\mathrm{arc}(\theta_2 - \theta_1)}{2\pi\sqrt{A^2 + B^2/2}} \tag{4.17}$$

where $\mathrm{arc}(\theta_2 - \theta_1)$ is the length of the contact arc and the denominator is an approximate formula for the perimeter of an ellipse, expressed in terms of the horizontal and vertical amplitudes A and B, respectively. The numerator in Eq. (4.17) must be calculated numerically for each specific ellipse geometry and HSR. The angular position of tool entry θ_1 is established from HSR and the ellipse dimensions A and B. The exit angular position θ_2 is determined by the depth of cut d and the ellipse dimensions. When $d/B \geq 1$, $\theta_2 = \pi$ and so $d/B = 1$ is established as the maximum value for the duty cycle at each HSR. The maximum value for duty cycle in 2D VAM is 0.5, which occurs when the value of F_{UP} is large enough so that successive elliptical cycles do not overlap, and when the depth of cut is equal to or greater than the ellipse vertical amplitude B.

Figure 4.6 gives a plot of the 2D VAM duty cycle as a function of HSR and d/B for a tool ellipse with $A = 11\,\mu\mathrm{m}$ and $B = 2\,\mu\mathrm{m}$.

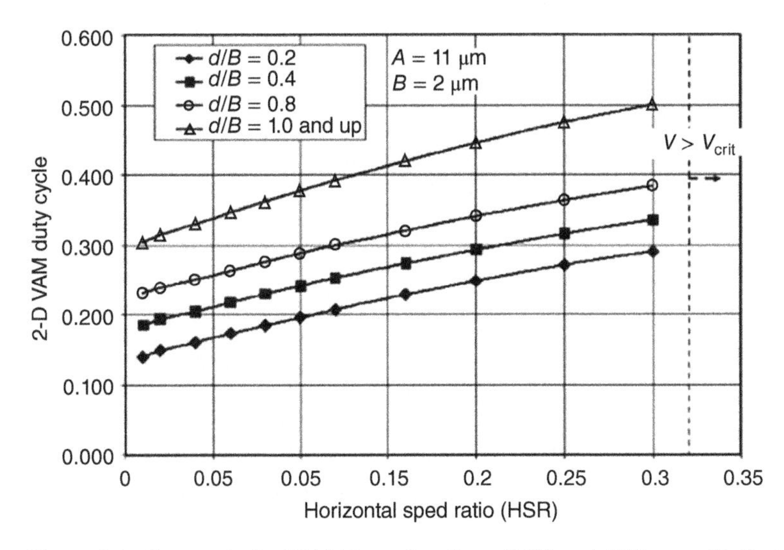

Figure 4.6 Duty cycle for 2D VAM as a function of HSR and *d/B* for an elliptical tool path with $A = 11\,\mu\mathrm{m}$ and $B = 2\,\mu\mathrm{m}$. Source: Cerniway [2].

4.3 Kinematics of Vibration-Assisted Milling

In VAMILL, vibration with high frequency and small amplitude can be superimposed onto the motion of either the tool or the workpiece. According to the dimensions of the vibration applied, VAMILL can also be divided into two groups: 1D and 2D VAMILL. In 1D VAMILL, vibration is applied either in the feed direction, i.e. feed-directional vibration-assisted milling (FVAMILL), or in the cross-feed direction, i.e. cross-feed-directional vibration-assisted milling (CFVAMILL). In 2D VAMILL, vibration is applied simultaneously in both feed and cross-feed directions (2DVAMILL).

Figure 4.7 shows a schematic diagram of VAMILL, where the workpiece feed, cross-feed, and axial depth of cut are in x-, y-, and z-directions, respectively. The tool tip motion without vibration imposed could be depicted by Eq. (4.18):

$$\begin{cases} x = r\sin\left[\omega t - \dfrac{2\pi(z_i - 1)}{Z}\right] \\ y = r\cos\left[\omega t - \dfrac{2\pi(z_i - 1)}{Z}\right] \end{cases} \tag{4.18}$$

where r and ω are the radius and angular velocity of the cutter, z_i is the ith cutter tooth, and Z is the number of flutes.

When a simple harmonic vibration is imposed onto the workpiece, the vibration trajec tory is:

$$\begin{cases} x_w = ft + A\sin(2\pi f_x t + \phi_x) \\ y_w = B\sin(2\pi f_y t + \phi_y) \end{cases} \tag{4.19}$$

where f is the feed velocity, A and B are the vibration amplitudes, f_x and f_y are the vibration frequencies, ϕ_x and ϕ_y are the phase angles, in the x- and y-directions. Then, the relative

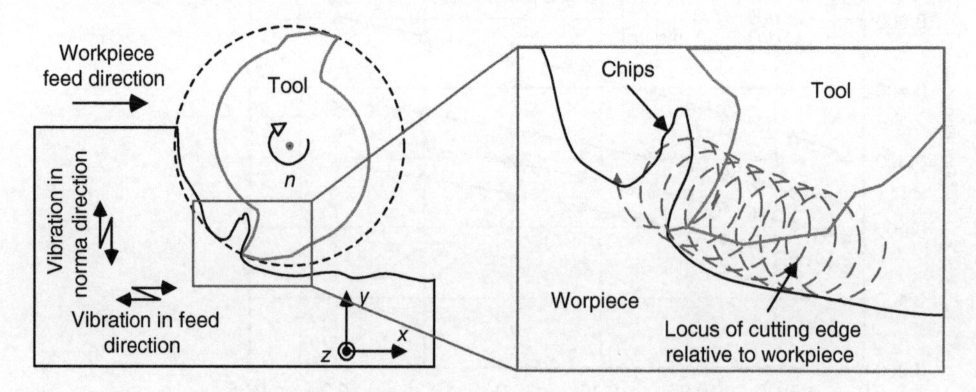

Figure 4.7 Schematic diagram of vibration-assisted milling. Source: Chen et al. [9].

displacement (x_i, y_i) between tool tip and workpiece can be expressed as follows:

$$\begin{cases} x_i = ft + r\sin\left[\omega t - \dfrac{2\pi(z_i - 1)}{Z}\right] + A\sin(f_x t + \phi_x) \\ y_i = r\cos\left[\omega t - \dfrac{2\pi(z_i - 1)}{Z}\right] + B\sin(f_y t + \phi_y) \end{cases} \begin{pmatrix} \text{CFVAMILL} : A = 0, B \neq 0 \\ \text{FVAMILL} : A \neq 0, B = 0 \\ \text{2DVAMILL} : A \neq 0, B \neq 0 \end{pmatrix} \tag{4.20}$$

4.3.1 Types of TWS in VAMilling

In VAM, the TWS refers to the rapid periodic interruption of constant tool workpiece contact. An appropriate pattern of TWS is the key to the success of VAMILL. Three different types of TWS mechanisms and their requirements are identified and described below.

4.3.1.1 Type I

Figure 4.8 illustrates the type I TWS in VAMILL. Type I separation occurs in the current tool path when the component of the relative velocity between tool and workpiece in the cutting direction (i.e. tangential component, V_t) is opposite to the tool rotation direction. This causes the tool tip to lag behind the workpiece so that separation occurs.

As shown in Figure 4.8, at the position 1, V_t would be greater than zero, and the tool is in contact with the workpiece. With the advance of the cutting tool, it would decrease to be equal to zero at position 2, when the tool would break contact with workpiece. Furthermore, at position 3, the component of the relative velocity in cutting direction has become negative, i.e. in the direction opposite to the tool rotation, and the tool separates from the workpiece. Thus, during the period of time when the tool and workpiece lose contact, V_t changes

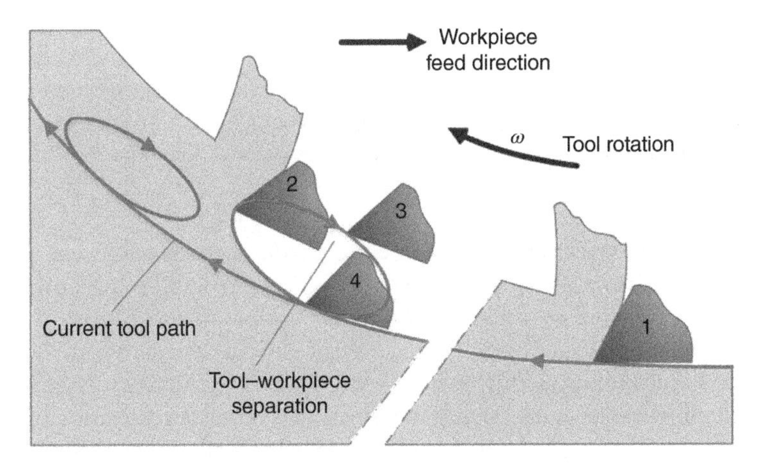

Figure 4.8 Type I TWS in VAMILL.

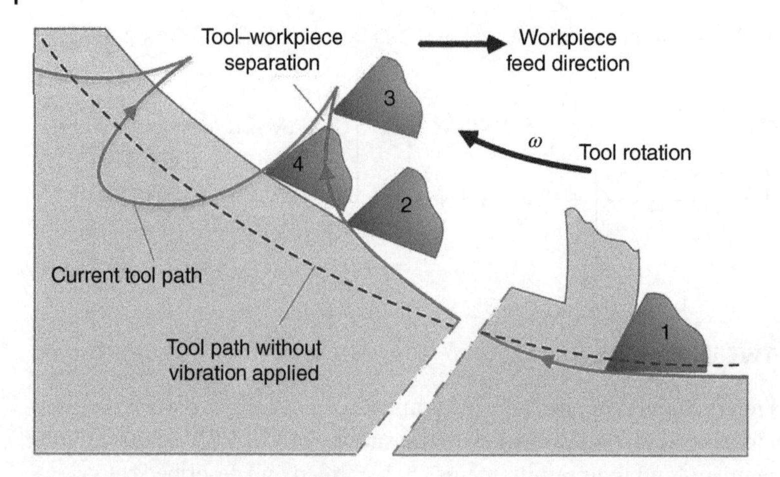

Figure 4.9 Type II TWS in VAMILL.

from zero to a negative value and then to a positive value. At position 4, the tool would make contact with the workpiece again. This type of TWS is similar to vibration-assisted turning.

4.3.1.2 Type II

Figure 4.9 shows a systematic diagram for type II TWS in VAMILL. Such type II separation occurs in the current tool path when vibration displacement in the instantaneous UCT direction (i.e. tool radial direction) is larger than instantaneous UCT. This would make the tool instantaneously break away from the workpiece, leading to the occurrence of tool-workpiece separation, namely, TWS.

As shown in Figure 4.9, when the vibration displacement in the tool radial direction is smaller than the instantaneous UCT, the cutting tool is in close contact with the workpiece surface (see position 1). Once the vibration displacement in the tool radial direction is equal to the instantaneous UCT, the actual instantaneous UCT will be zero at position 2. This means that the cutting tool is about to break contact with workpiece. Then, at position 3, the vibration displacement will exceed the instantaneous UCT, and the tool remains separated from the workpiece. Finally, when the vibration displacement is equal to the instantaneous UCT at position 4, the tool regains contact with the workpiece.

4.3.1.3 Type III

In type I and II separation, the effect of the uneven cutting path in the previous cut is ignored, but it should be considered when the depth of cut is comparable with the vibration amplitude. Figure 4.10 illustrates a schematic diagram for type III TWS in VAMILL. It can be seen that with the assistance of vibration, the current tool path overlaps with the surface contour left by previous cutting path(s) at some regions, where the cutting tool edge may break contact with the workpiece and discontinuous chips are generated. Because part of the materials in the current cutting path has been removed by previous cutting path(s), periodical TWS takes place.

In addition, it should be noted that in VAMILL, type I, II, and III separations could always happen simultaneously. In specific conditions, it is kinematically possible to obtain only a certain type of separation, which will be discussed in Section 4.3.2.

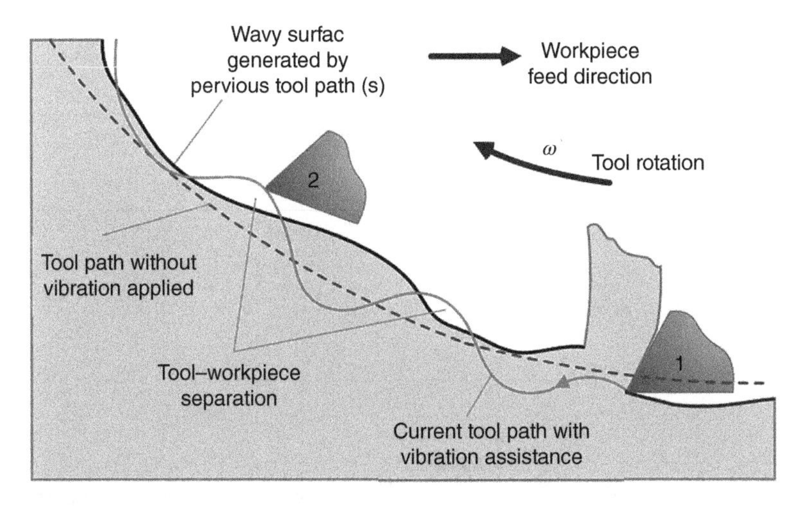

Figure 4.10 Type III TWS in VAMILL.

4.3.2 Requirements of TWS

In this section, the requirements for three types of separation to occur are discussed. In high-speed milling, because the feeding speed is much smaller than the tool rotation and high frequency vibration speed, the workpiece feed rate effect is usually negligible. This section only focuses on the separation requirements for 1D VAMILL.

4.3.2.1 Type I Separation Requirements

The occurrence of type I separation is due to the relative velocity between the tool tip and workpiece in the cutting direction, or the nominal cutting velocity, V_t. Without vibration assistance, this can be expressed as:

$$V_t = \omega r \tag{4.21}$$

Similarly, for the instantaneous cutting direction, the velocity components in the x and y directions of the workpiece could be obtained through differentiating Eq. (4.19) as:

$$\begin{cases} V_x(t) = f + 2\pi A f_x \cos(2\pi f_x t + \phi_x) \\ V_y(t) = 2\pi B f_y \cos(2\pi f_y t + \phi_y) \end{cases} \tag{4.22}$$

(1) ***Vibration in the cross-feed direction (CFVA milling)***
When the vibration is applied in the cross-feed direction, where $A = 0$, the tool center orbit turns into a harmonic linear locus along the cross-feed direction. The following requirements for type I separation should be met:

$$2\pi B f_y \cos(2\pi f_y t + \phi_y) \sin\theta \geq \omega r \tag{4.23}$$

where θ is the tool rotation angle at time t. Rearrange Eq. (4.23):

$$\cos(2\pi f_y t + \phi_y) \sin\theta \geq \frac{\omega r}{2\pi B f_y} \tag{4.24}$$

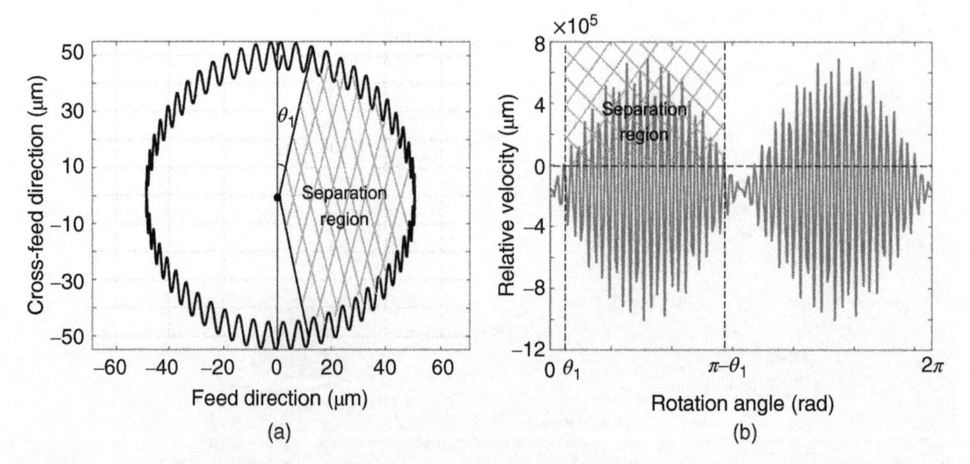

Figure 4.11 (a) The region where type 1 separation is likely to occur during one cycle of tool path and (b) the relative velocity in cutting direction. Source: Chen et al. [9].

Obviously, Eq. (4.24) is solvable when

$$\frac{\omega r}{2\pi B f_y} < 1 \tag{4.25}$$

For instance, as shown in Figure 4.11, within a full circle of tool rotation, type I separation is likely to occur in the region when:

$$\theta_1 < \theta < \pi - \theta_1 \tag{4.26}$$

where $\theta = \theta_1$ is the solution of Eqs. (4.23–4.25)

When the tool rotation angle is in the range $[\theta_1, \pi - \theta_1]$, TWS takes place and the tool path is as shown in Figure 4.11a. In Figure 4.11b, the instantaneous cutting speed during a whole circle of tool rotation is numerically calculated.

From Eq. (4.24), it can be found that the separation zone will increase with an increase in maximum vibration velocity, $2\pi B f_y$, and a decrease in the nominal cutting velocity, ωr. When $\omega r > 2\pi B f_y$, Eq. (4.24) becomes unsolvable and no type I separation would occur in the whole cycle.

(2) ***Vibration in the feed direction (FVA milling)***

When the vibration is applied in the feed direction, where $B = 0$, the tool center performs a 1D sinusoidal vibration along the feed direction. As with the case in Section 4.3.1.1, the requirement for type I separation is

$$2\pi A f_x \cos(2\pi f_x t + \phi_x) \cos\theta \geq \omega r \tag{4.27}$$

Rearrange Eq. (4.27)

$$\cos(2\pi f_x t + \phi_x) \cos\theta \geq \frac{\omega r}{2\pi A f_x} \tag{4.28}$$

Equation (4.28) is solvable when

$$\frac{\omega r}{2\pi A f_x} > 1 \tag{4.29}$$

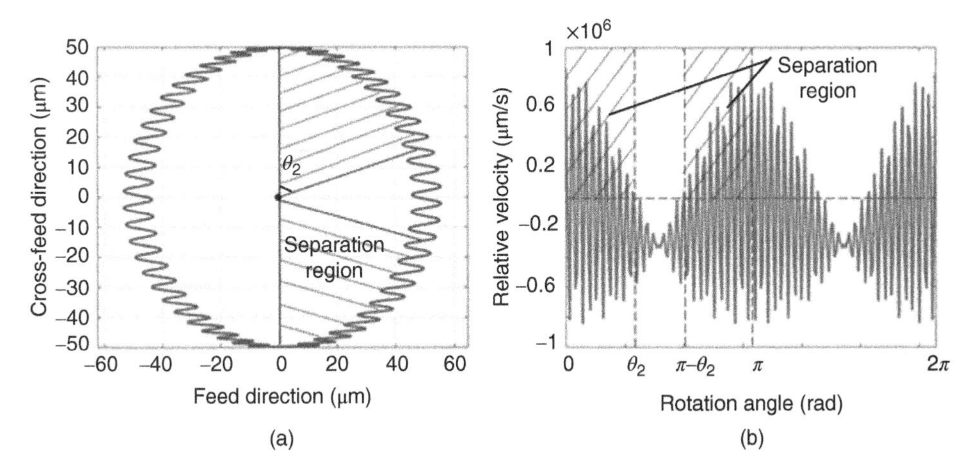

Figure 4.12 (a) The region where type 1 separation is likely to occur during one cycle of tool path and (b) relative velocity in cutting direction. Source: Chen et al. [9].

Figure 4.12 shows the kinematic simulation result for one full tool rotation cycle in FVA milling. When θ is smaller than θ_2, but larger than $\pi - \theta_2$, TWS takes place as shown in Figure 4.12a. On this basis, the instantaneous cutting speed during a whole circle of tool rotation is numerically calculated (see Figure 4.12b).

As depicted above, Eq. (4.28) indicates that the range of the separation increases with the increase of the maximum vibration velocity, $2\pi A f_x$, and the decrease in the nominal cutting velocity, ωr. When the maximum vibration velocity is lower than the nominal cutting velocity, Eq. (4.28) becomes $\cos(2\pi f_x t + \phi_x)\cos\ \theta > 1$, which is unsolvable, and no type I separation would occur in the whole cycle.

4.3.2.2 Type II Separation Requirements

Type II separation depends on the relative displacement between the nominal UCT and the vibration displacement. In the conventional milling process, the instantaneous UCT, h_D, can be expressed by

$$h_D = f_z \sin\ \theta \tag{4.30}$$

where f_z is feed per tooth.

(1) ***Vibration in the cross-feed direction (CFVA milling)***

When the vibration is imposed in cross-feed direction, as shown in Figure 4.13a, the instantaneous UCT, h_{DV}, can be expressed by

$$h_{DV} = f_z \sin\ \theta - y_w \cos\ \theta \tag{4.31}$$

Based on Eq. (4.31), it can be noted that when $\tan\theta < \frac{y_w}{f_z}$, the UCT is smaller than zero; thus, type II separation could occur as shown in Figure 4.13b. Moreover, the separation region increases with the increase in y_w. However, the periodic separation cannot be achieved over the full cutting path, as shown in Figure 4.13b.

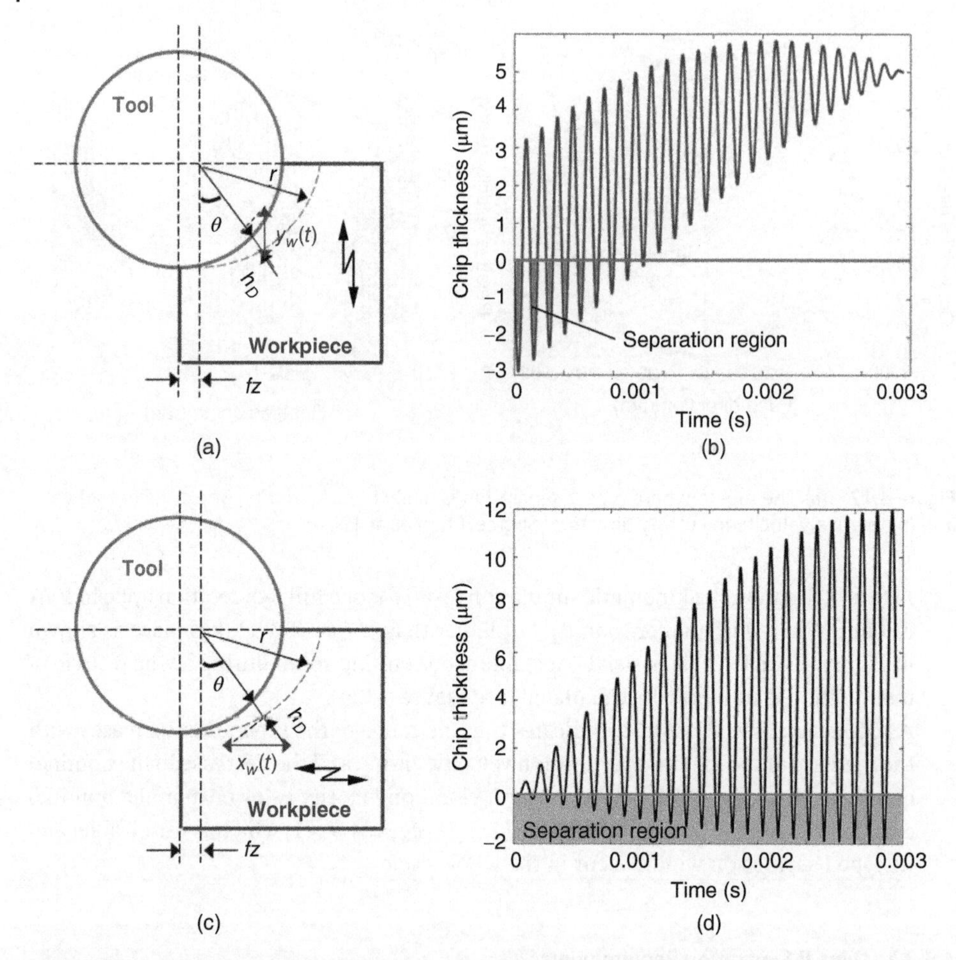

Figure 4.13 Type II separation. (a, b) Schematic diagram and instantaneous uncut chip thickness in CFVAMILL and (c, d) schematic diagram and instantaneous uncut chip thickness in FVAMILL. Source: Chen et al. [9].

(2) ***Vibration in the feed direction (FVA milling)***

When vibration is applied in the feed direction, as shown in Figure 4.13c, the instantaneous UCT can be expressed by

$$h_{DV} = f_z \sin\ \theta - x_w \sin\ \theta = (f_z - x_w) \sin\ \theta \tag{4.32}$$

When $f_z - x_w < 0$, the UCT would become less than zero and type II separation will occur, as shown in Figure 4.13d. To achieve periodical type II separation, the vibration should be applied in the feed direction, and the vibration amplitude should be larger than the feed per tooth. However, in such a case, a large vibration amplitude will increase the tool wear, leading to the deterioration of the surface quality of the machined surface. Therefore, such kinds of periodical type II separation mechanisms should be eliminated.

4.3.2.3 Type III Separation Requirements

As for type III separation, it is dependent on the surface contour generated by previous tool paths. It is difficult to study the underlying separation mechanism using a mathematical expression. For this reason, the numerical simulation method is adopted to determine the requirements for type III separation. The simulation was conducted using specific parameter sets in micro-milling, but the findings can be generalized. The requirements for two types of type III separation should also be analyzed, depending on the vibration direction used.

(1) ***Vibration in cross-feed direction***

In this section, a two-flute end mills with a diameter of 0.1 mm and a rotation speed of 5000 rpm are employed in the simulations. The value of feed per tooth and vibration amplitude are set at 6 and 4 μm, respectively. Two typical vibration frequencies, representing odd and even multiples of the rotation speed, are applied in the cross-feed direction to investigate their influence on the TWS.

Figure 4.14 shows the simulation results for the tool tip trajectory and UCT when the values of vibration frequency are 71 and 72 times the spindle rotation frequency. The vibration in the cross-feed direction can change the trajectory of the tool tip. It induces the constant fluctuation of the UCT but does not cause the periodic TWS over whole tool path. Although the vibration increases the tool tip displacement in cross-feed direction, the tool and workpiece can be separated in the cutting in and cutting out regions, as circled in Figure 4.14b,d. As the tool tip moves in the feed direction, the vibration displacement in the direction of UCT would obviously decrease, and so the separation time period of the tool and the workpiece decreases until TWS disappears completely.

In summary, the vibration applied in the cross-feed direction has imposed different effects on the trajectory of the cutting tool, depending on the vibration frequency:

When the applied vibration frequency is an odd multiple of the spindle rotation frequency, the vibration significantly increases the maximum cutting thickness, as shown in Figure 4.14b, which leads to worsening of the machined surface roughness.

However, when the vibration frequency is an even multiple of the spindle rotation frequency, the vibration does not change the maximum cutting thickness, and the average cutting forces are expected to be reduced. Regardless of the relationship between the vibration frequency and the spindle rotation frequency, a larger vibration amplitude is needed to ensure TWS. In such a case, there is a lager variation in the separation time period and a poor regularity. As a result, this will cause a larger self-excited vibration between the tool and workpiece, which might worsen the machined surface quality and reduce machining accuracy and tool life.

Therefore, the vibration applied in the cross-feed direction could reduce the average cutting force at the appropriate vibration parameters, but it cannot achieve periodic separation, and hence, it is not suitable for use alone.

(2) ***Vibration in feed direction***

Figure 4.15 shows the tool tip trajectory and UCT when the applied vibration frequencies are odd and even multiples of the spindle rotation frequency. It can be seen from Figure 4.15a,d that the vibration in the feed direction changes the tool tip trajectory, leading to an UCT, which fluctuates constantly.

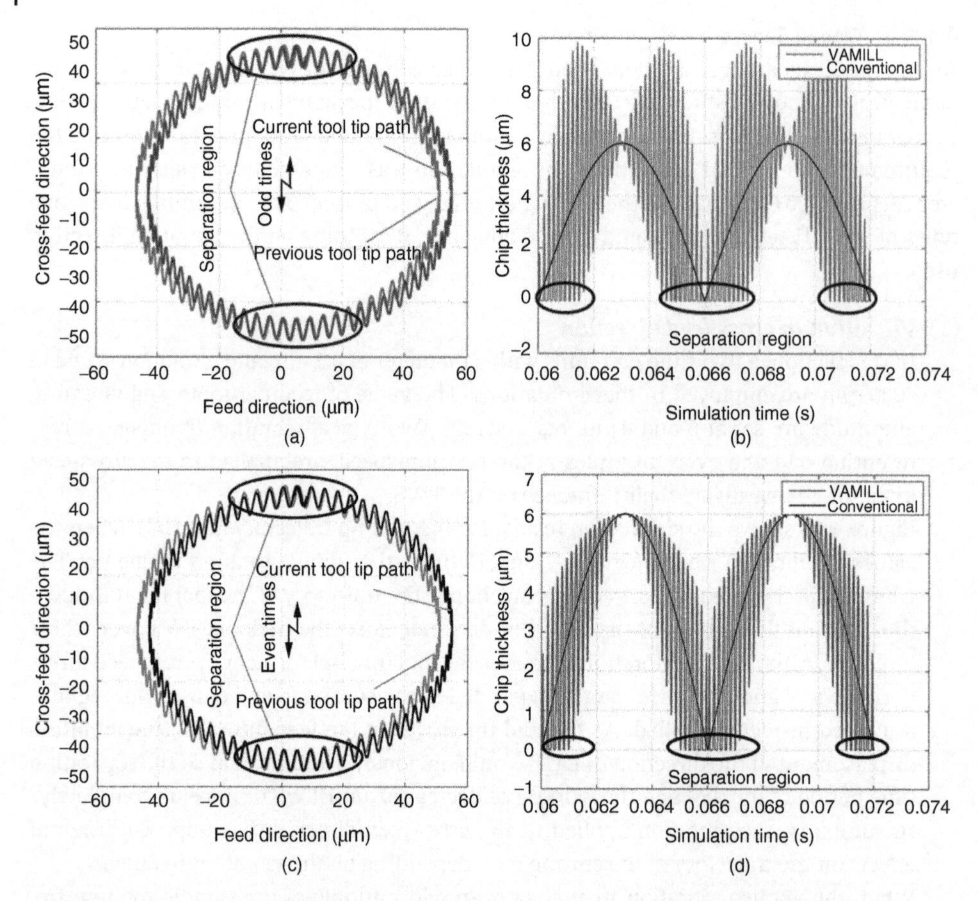

Figure 4.14 Type III separation in cross-feed direction. (a, b) The tool tip trajectory and instantaneous uncut chip thickness at the vibration frequency of odd times of spindle speed. (c, d) The tool tip trajectory and instantaneous uncut chip thickness at the vibration frequency of even times of the spindle speed. Source: Chen et al. [9].

When the vibration frequency applied is an odd multiple of the spindle rotation frequency, the peaks (troughs) of waves of the ith tool tip trajectory overlap with the troughs (peaks) of waves of the $(i+1)$th tool tip trajectory, as shown in Figure 4.15a. Thus, if the vibration amplitude is larger than half of the feed per tooth, the periodical separation of tool tip and the workpiece is regular during the whole cutting process as shown in Figure 4.15b. However, when the vibration frequency is an even multiple of the spindle rotation frequency, the peaks (troughs) of waves in the ith tool tip trajectory would overlap with the peaks (troughs) of wave in the $(i+1)$th tool tip trajectory as illustrated in Figure 4.15c. This can lead to the constant fluctuation of the UCT fluctuate, but no type III separation occurs during the whole cutting process as shown in Figure 4.15d.

In summary, at a vibration frequency of an even multiple of the spindle rotation frequency, the vibration amplitude is greater than half of the feed per tooth, and periodic type III separation can be achieved to improve the machined surface roughness.

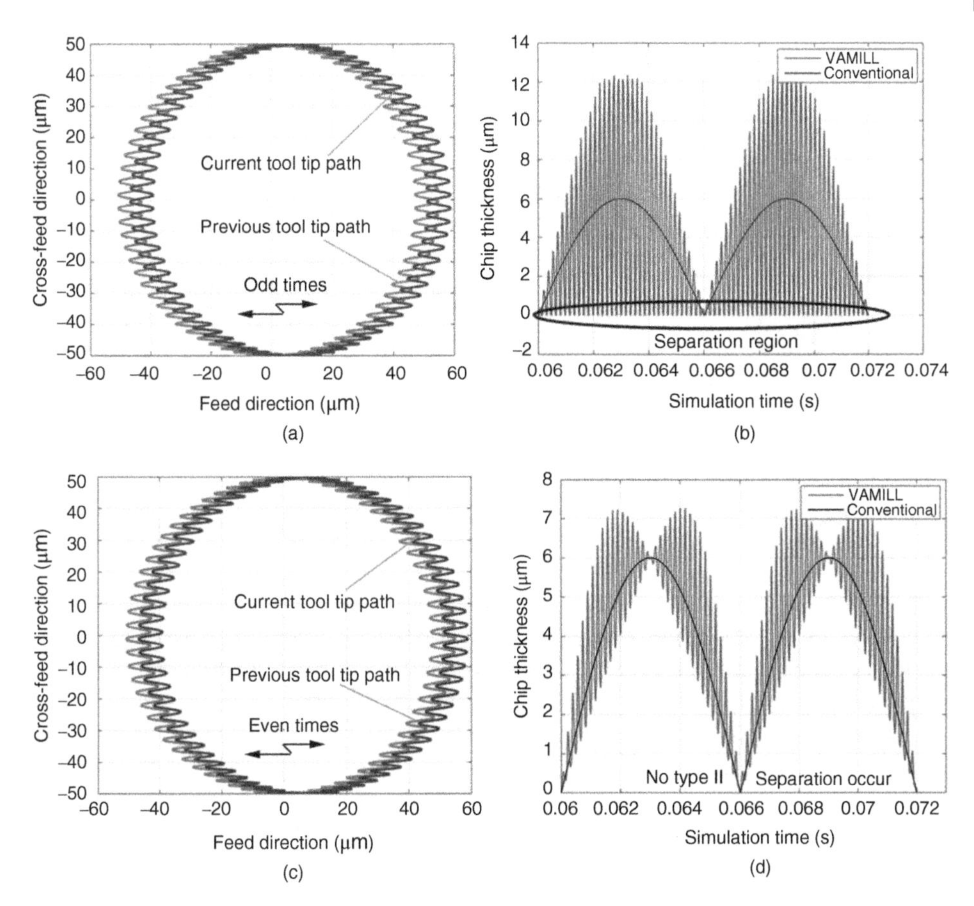

Figure 4.15 Type III separation in feed direction. (a, b) The trajectory of the tool tip and instantaneous uncut chip thickness when the vibration frequency is odd times of the spindle speed. (c, d) The trajectory of the tool tip and instantaneous uncut chip thickness when the vibration frequency is even times of the spindle speed. Source: Chen et al. [9].

However, when the applied vibration frequency is an odd multiple of the spindle rotation frequency, the vibration will increase the maximum UCT, thereby worsening the machining quality.

Thus, the vibration applied in the feed direction should be preferably adopted to realize periodic type III separation rather than that in cross-feed direction.

4.4 Finite Element Simulation of Vibration-Assisted Milling

Because of the high frequency and small amplitude of the vibration applied in VAMILL, the separation time is always less than 10^{-5} seconds, and it is difficult to observe TWS experimentally. Finite element (FE) modeling has been proven to be an effective method to investigate the cutting process, particularly the hard-to-observe cutting phenomena of the chip formation [4, 5], minimum chip thickness [6], and temperature in the cutting zone [7].

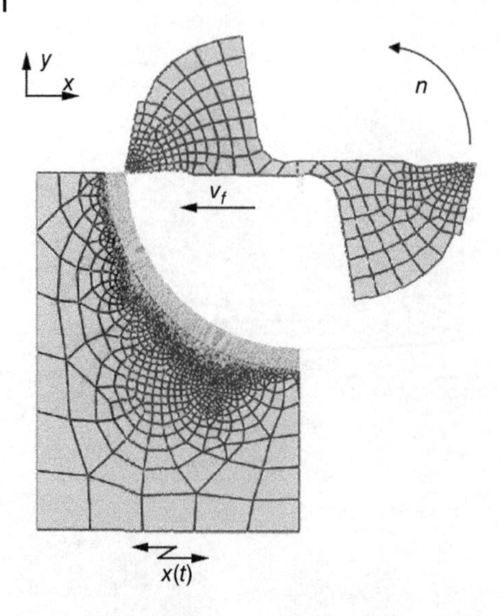

Figure 4.16 FE model of VAMILL.

In this study, a FE model was established using the commercial software package ABAQUS/Explicit to support the proposed kinematic model and three types of TWS, as shown in Figure 4.16. Here, AISI 1045 steel is chosen as the workpiece material because of its popularity in plastic injection molding industry. The cutter is set up as a rigid body in order to increase the simulation speed and to exclude the effect of tool deformation on the cutting process. The cutting-edge radius and minor cutting-edge angle of the tool are 3 μm and 5°, respectively. The nonlinear temperature and strain rate-sensitive Johnson–Cook (JC) material model and Johnson–Cook damage model are used to describe the workpiece material properties.

The primary equation of the JC model describes the flow stress as:

$$\sigma_y = [A + B(\varepsilon_p)^n][1 + Cln(\dot{\varepsilon}_p^*)][1 - (T^*)^m] \tag{4.33}$$

where $\dot{\varepsilon}_p^* = \frac{\dot{\varepsilon}_p}{\dot{\varepsilon}_{p0}}$, $T^* = \frac{T-T_0}{T_m-T_0}$, and ε_p is the effective plastic strain, $\dot{\varepsilon}_p$and $\dot{\varepsilon}_{p0}$ are the plastic strain rate and effective plastic strain rate used for calibration of the model, respectively, and T and T_0 are the current and reference temperatures, respectively. The parameters A, B, n, C, m, and T_m along with other parameters are extracted from previous research [8].

The machining parameters used in the FE simulations are summarized in Table 4.2. They are selected based on the proposed requirements for each type of TWS, in order to verify the correctness and reliability of the proposed kinematic models.

As discussed above, the vibration applied in the feed direction is preferable to realize the type II and type III TWS, and thus, the vibration is applied in the feed direction in the FE simulations No. 2 and No. 3, while the vibration is applied in the cross-feed direction in the simulation No. 1.

Figures 4.17–4.19 illustrate the chip formation process at various stages for simulation No. 1–3. When the vibration speed of the workpiece (v_{wf}) in the cutting direction is less than the linear speed of the tool tip due to tool rotation (ωr), the workpiece materials are removed from the workpiece, as shown in Figure 4.17a. With the increase of v_{wf} to be equal to ωr, there is no relative motion between tool and workpiece materials. TWS is about to

Table 4.2 Machining parameters used in the FE simulations.

FE simulation No.	Spindle speed (rpm)	Feed per tooth (μm)	Vibration amplitude (μm)	Vibration frequency (× spindle frequency)
1	5000	6	4	71
2	5000	6	6	71
3	5000	6	4	27

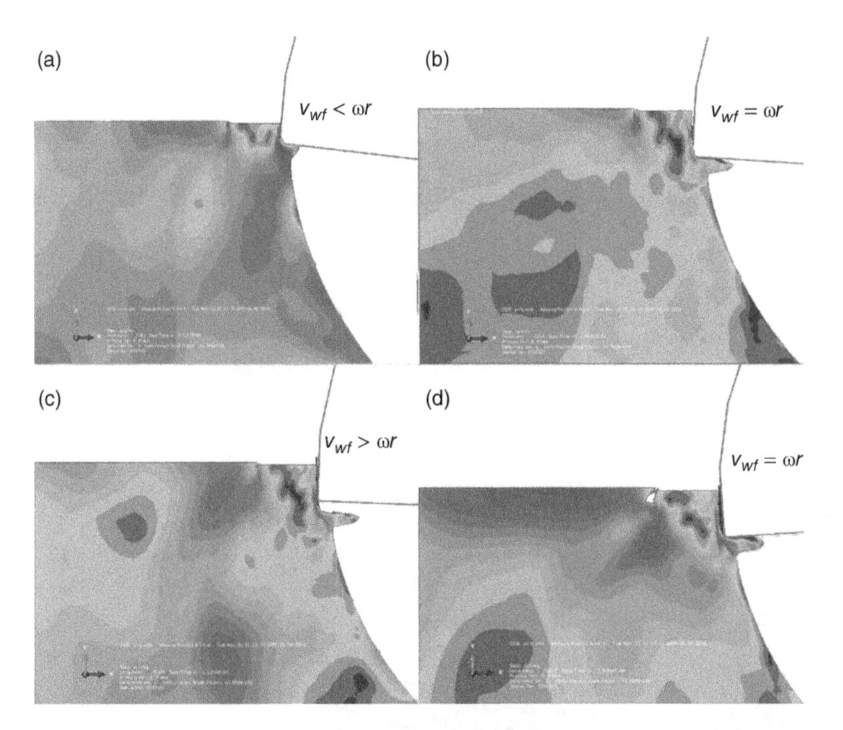

Figure 4.17 Type I TWS in VAMILL in FE simulation No. 1.

begin, as displayed in Figure 4.17b. Until v_{wf} is greater than ωr, the tool tip lags behind the workpiece and type I TWS takes place (see Figure 4.17c). Finally, v_{wf} is decreased to be equaling to ωr, the tool regains contact with the workpiece, and the cycle repeats. The simulation results follow the proposed mechanism of the type I TWS in VAMILL.

When the vibration is applied in the feed direction, type II TWS takes place, depending on the relationship between the instantaneous vibration displacement $x(t)$ of the workpiece in the feed direction and instantaneous UCT (h_D), as given in Figure 4.18. When $x(t)$ is smaller than h_D, the workpiece material is successfully removed under the action of cutting tool. When $x(t)$ is equal to or larger than h_D, the tool loses contact with the workpiece, leading to the occurrence of type II TWS.

Figure 4.19 shows periodic separation because of the overlap of the current and previous cutting paths, which is obtained in FE simulation No. 3. Here, in order to clearly illustrate

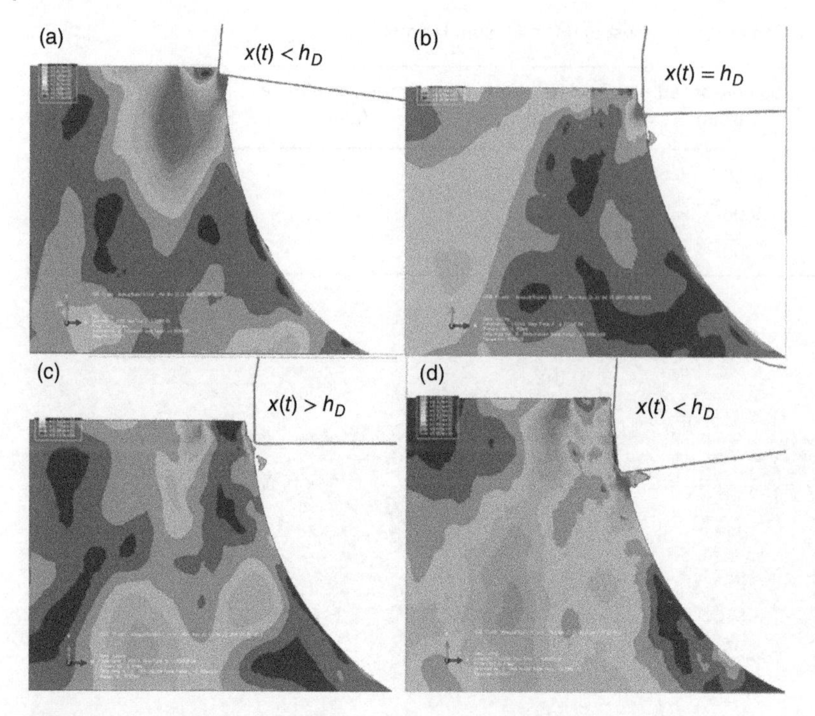

Figure 4.18 Type II TWS in VAMILL in FE simulation No. 2.

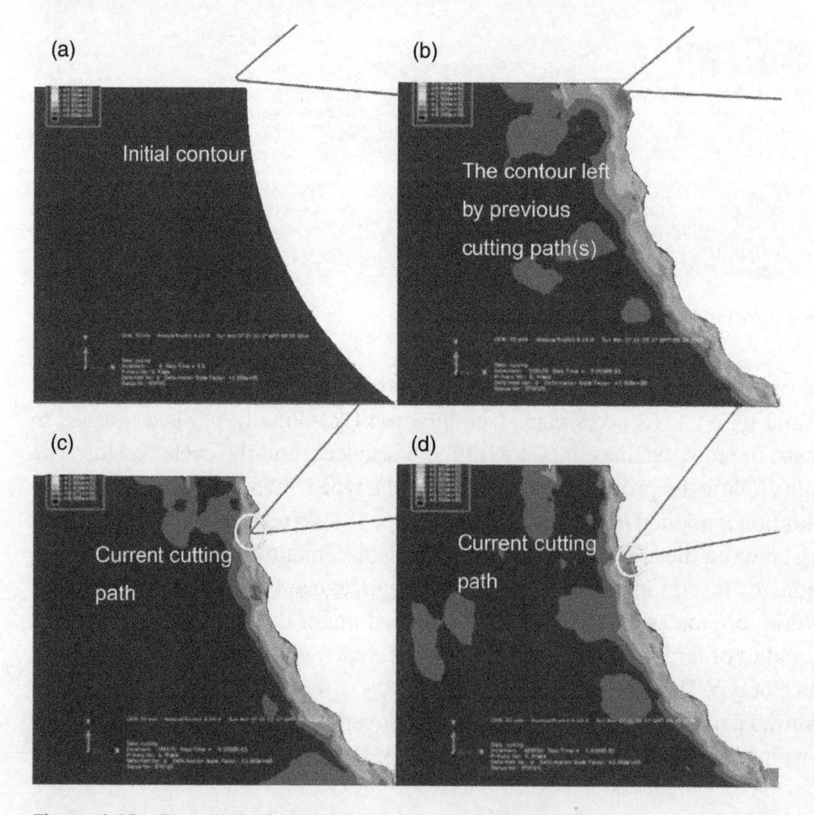

Figure 4.19 Type III TWS in VAMILL in FE simulation No. 3.

the TWS, a much lower vibration frequency is used in simulation. The original surface and processed surface obtained by previous cutting path are shown in Figure 4.19a,b. When the workpiece vibration is applied in the feed direction, it can be seen in Figure 4.19c,d that, because the new cut profile is not consistent with the previous cut profile, the cutter trajectories overlap. Consequently, the tool will break away from the workpiece surface, causing periodic type III separation.

Therefore, the simulation results verify the correctness and reliability of the separation requirements obtained from the kinematic analysis. Each type of TWS could be realized by using the machining parameters given by the kinematic models.

4.5 Conclusion

The kinematics of VAM are studied, and the TWS type and the requirements for each separation type are discussed. The formulated kinematics and separation requirements are of great significance to determine the optimal machining and vibration parameters and to provide the technological principles for the design of VAM system.

References

1 Xu, W.X. and Zhang, L.C. (2015). Ultrasonic vibration-assisted machining: principle, design and application. *Adv. Manuf.* 3 (3): 173–192.

2 Cerniway, M.A. (2001). Elliptical diamond milling: kinematics, force, and tool wear. MS thesis. North Carolina State University.

3 Klocke, F., Demmer, A., and Heselhaus, M. (2004). Material removal mechanisms in ultrasonic-assisted diamond turning of brittle materials. *Int. J. Mater. Prod. Technol.* 20: 231–238.

4 Özel, T., Thepsonthi, T., Ulutan, D. et al. (2011). Experiments and finite element simulations on micro-milling of Ti–6Al–4V alloy with uncoated and cBN coated micro-tools. *CIRP Ann. Manuf. Technol.* 60 (1): 85–88.

5 Zhang, B. and Bagchi, A. (1994). Finite element simulation of chip formation and comparison with machining experiment. *Trans. ASME J Eng. Ind.* 116: 289–289.

6 Lai, X. et al. (2008). Modelling and analysis of micro scale milling considering size effect, micro cutter edge radius and minimum chip thickness. *Int. J. Mach. Tools Manuf.* 48 (1): 1–14.

7 Wu, H.B. and Zhang, S.J. (2014). 3D FEM simulation of milling process for titanium alloy Ti6Al4V. *Int. J. Adv. Manuf. Technol.* 71 (5–8): 1319–1326.

8 Simoneau, A., Ng, E., and Elbestawi, M. (2007). Grain size and orientation effects when microcutting AISI 1045 steel. *CIRP Ann. Manuf. Technol.* 56: 57–60.

9 Chen, W., Huo, D., Hale, J., and Ding, H. (2018). Kinematics and tool-workpiece separation analysis of vibration assisted milling. *Int. J. Mech. Sci.* 136: 169–178.

5

Tool Wear and Burr Formation Analysis in Vibration-Assisted Machining

5.1 Introduction

Increasing tool life and part quality is imperative in high-value machining performance and productivity. The reduction of tool wear and suppression of burr formation are the main benefits of applying vibration assistance to the machining process; the former is more relevant with hard-to-machine materials and the latter is applicable to all ductile materials. Numerous research efforts have focused on investigating the removal and deformation mechanisms of materials in vibration-assisted machining from the perspective of tool wear and burr formation. This chapter considers the mechanisms of the tool wear and burr formation and their classification, and reviews current developments in tool wear reduction and burr formation suppression through vibration-assisted machining.

5.2 Tool Wear

Cutting tools are subjected to extremely severe friction due to the metal-to-metal contact between the cutting tool and workpiece during material processing, which causes high level of stress and temperature near the surface of the tool and hence leads to unavoidable tool wear. Worn tools can reduce machining accuracy and surface finish, which in turn leads to deteriorating cutting efficiency and increased machining costs. Cutting parameter optimization is a common method for tool wear reduction and extending tool life in conventional machining, but its effect is limited. This section introduces the effect of vibration-assisted machining in reducing tool wear. The effect of vibration-assisted machining technology on tool wear can be obtained by analyzing the mechanisms and appearance of tool wear, as well as the characteristics of vibration-assisted machining.

5.2.1 Classification of Tool Wear

Tool wear can be divided into two groups: abnormal and normal wear, as shown in Figure 5.1. Abnormal wear is usually caused by impact or uneven heating during machining process, leading to the brittle damage of cutting tools such as chipping, fracture, spalling, and cracking. Normal wear is a process in which tool wear gradually increases

Vibration Assisted Machining: Theory, Modelling and Applications,
First Edition. Lu Zheng, Wanqun Chen, and Dehong Huo.

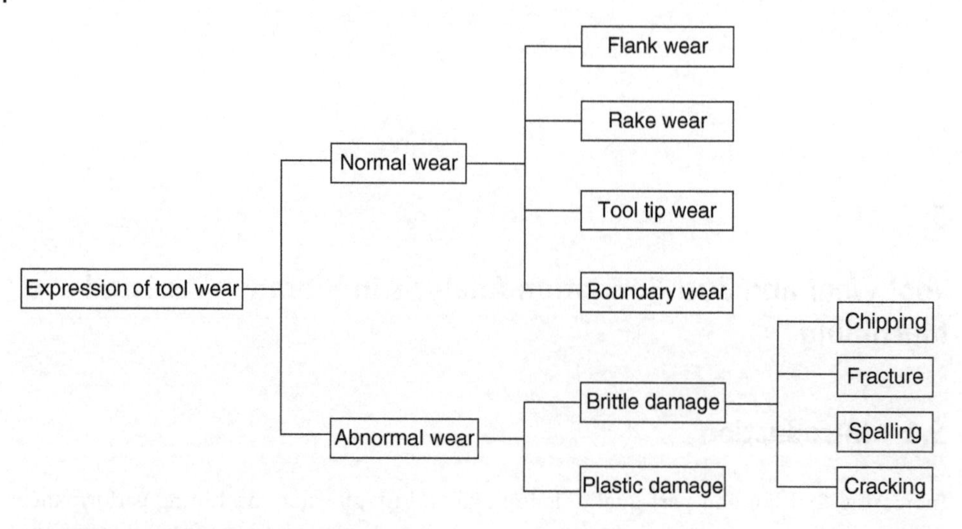

Figure 5.1 Classification of tool wear.

over time. It is caused by the friction and thermochemical effects during the machining process, leading to rake wear, flank wear, tool tip wear, and boundary wear.

Rake wear usually occurs when processing plastic materials, leading to a crater behind the cutting edge on the tool rake face due to high cutting speeds, as shown in Figure 5.2. The position on the tool rake face that reaches the highest temperature is usually where the wear starts, and the area and depth of the crater will expend and increase over time, leading to the increased rake angle of the cutting tool, which in turn improves the cutting conditions and makes it easy for the chips to curl and break. However, the strength of the cutting tool will be weakened as the crater continues to grow, leading to the breakage and chipping of cutting tools.

Flank wear is caused by the strong friction between the machined surface and the tool flank face, and a uniform area with a zero clearance angle on the tool flank face eventually forms. This usually happens along with tool tip wear and boundary wear, as shown in Figure 5.2. In addition, flank wear is usually used as the standard for tool wear since it can affect the dimensional accuracy and surface quality of the machined surface.

5.2.2 Wear Mechanism and Influencing Factors

Understanding the mechanisms and factors affecting tool wear during machining is very important in order to improve tool life and cutting performance. Mechanical and thermochemical wear mainly contribute to the normal wear process of cutting tools. Mechanical wear, including abrasive wear, is wear on the tool friction surface due to the intense friction occurring during processing. Meanwhile thermochemical wear is due to chemical reactions between the tool and the workpiece materials in high temperature conditions. Some wear mechanisms only occur in specific tool materials when processing certain workpiece materials, but multiple wear mechanisms usually work together during the tool wear process. These wear mechanisms include the following:

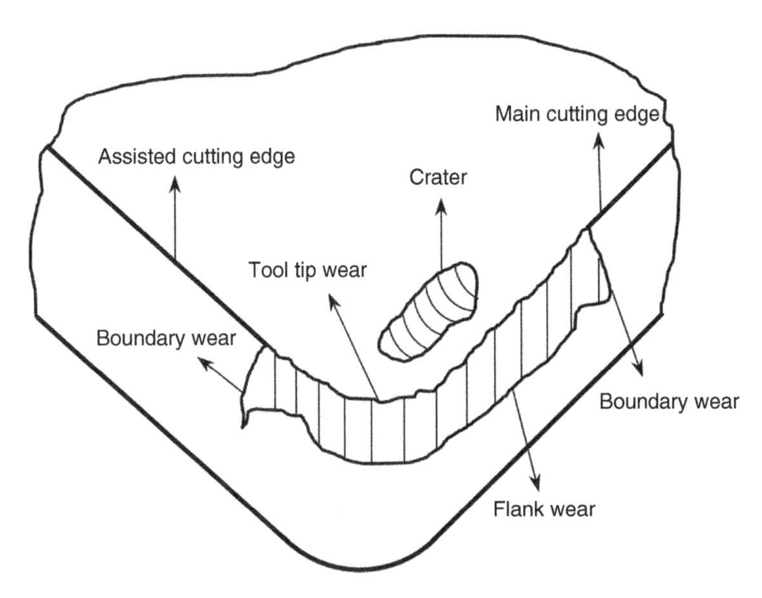

Figure 5.2 Layout of tool wear.

(1) **Coating delamination** only takes place on coated cutting tools and is believed to be the initial wear mechanism of the coated tool. A chemical reaction between the workpiece and coating material and crack propagation at the cutting tool substrate interface caused by differences in the thermal coefficient of expansion between the coating matrix and the tool substrate have been proposed as the main reasons for coating delamination. In addition, severe friction during processing also speeds up this phenomenon.
(2) **Abrasive wear** is one of the most common wear mechanisms and occurs in most metal cutting processes. It is caused by friction between the tool surface and the workpiece and leads to micro-grooves on the worn tool (especially on the tool rake face) due to the high-hardness particles contained in the workpiece. Abrasive wear can be reduced effectively using the cutting tools with higher material hardness.
(3) **Adhesion wear**, also called "cold welding," is caused by the adsorption of atoms between the friction surfaces of cutting tool and workpiece (chips) at specific pressure levels and high temperatures. The materials at the bonding points of the two friction surfaces (workpiece and cutting) become torn due to the relative motion in the machining process, and adhesion wear will occur when the particles on the tool cutting surface are taken away by the workpiece surface. The plasticity of the material to be processed and cutting conditions are the main factors affecting adhesion wear, and the effect of changing the cutting tool material used is quite limited in terms of reducing the adhesion wear.
(4) **Diffusion wear** is the result of a chemical reaction. Molecular activity between the tool and the workpiece contact surface is increased due to high temperature and pressure during processing, which leads to the diffusion and replacement of alloy elements such as titanium (Ti), tungsten (W), and cobalt (Co) from the cutting tool into the workpiece. Meanwhile, the element iron (Fe) in the workpiece diffuses into the cutting tool and reduces the mechanical properties of the tool material. As a consequence, diffusion

wear occurs due to the severe friction effect during the machining process. It is greatly affected by the cutting temperature and speed, and higher speed and temperature will speed up diffusion wear.

(5) **Oxidation wear** is also caused by the high temperature and pressure in the machining process. Unlike diffusion wear, oxidation wear requires the intervention of oxygen. An oxidation reaction on the tool surface generates a layer of brittle oxide, such as CoO and WO_2. As the machining process goes on, the brittle oxide on the tool surface is taken away by the workpiece or chip causing cutting tool wear. This usually occurs in the tool cutting edge area, which is in contact with the surrounding air.

Variables affecting tool wear mainly include the following four factors:

(1) *Workpiece materials.* Cutting force and cutting energy are highly dependent on the workpiece material and its physical and mechanical properties, including microstructure, hardness, and mechanical and thermal properties, which in turn affect tool wear. The plasticity, hardness, and thermal conductivity of the material have the greatest impact on tool wear.
(2) *Cutting parameters.* Parameters such as cutting speed, cutting depth, and feed rate have a great influence on the heat of cutting and cutting forces during machining, which in turn affect tool wear.
(3) *Cutting tool.* To obtain the optimum cutting performance, suitable tool parameters such as tool material, coating, and geometry need to be considered for different cutting conditions (roughing or finishing).
(4) *Machine tool.* The dynamic characteristics of the machine tool, which are determined by the machine structure, have a great influence on the stability of the cutting process. Unstable cutting processes are usually accompanied by large flutter, leading to a high fluctuating load on the tool. As a result, tool failure occurs due to the excessive wear or chipping of the cutting tool.

5.2.3 Tool Wear Reduction in Vibration-Assisted Machining

The life of a cutting tool can be significantly extended in vibration-assisted machining compared to the corresponding conventional machining, especially in the machining of hard and brittle materials. Generally, longer tool life can be achieved in 2D vibration-assisted machining compared with 1D vibration-assisted machining under the same cutting conditions. In order to fully understand the effect of vibration-assisted machining on tool wear, it is necessary to discuss the reduction of tool wear according to different mechanisms since different wear mechanisms only appear in particular combinations of specific tools and workpiece materials.

5.2.3.1 Mechanical Wear Suppression in 1D Vibration-Assisted Machining

Mechanical wear occurs irrespective of tool and workpiece material and combinations, and it appears to be the main cause of tool wear for ferrous metal processing, regardless of tool types. In addition, due to the large cutting force, the most severe mechanical wear occurs when machining hard and brittle ferrous materials such as hardened steel, nickel, or Inconel, which are also the most widely used in vibration-assisted machining. Figures 5.3

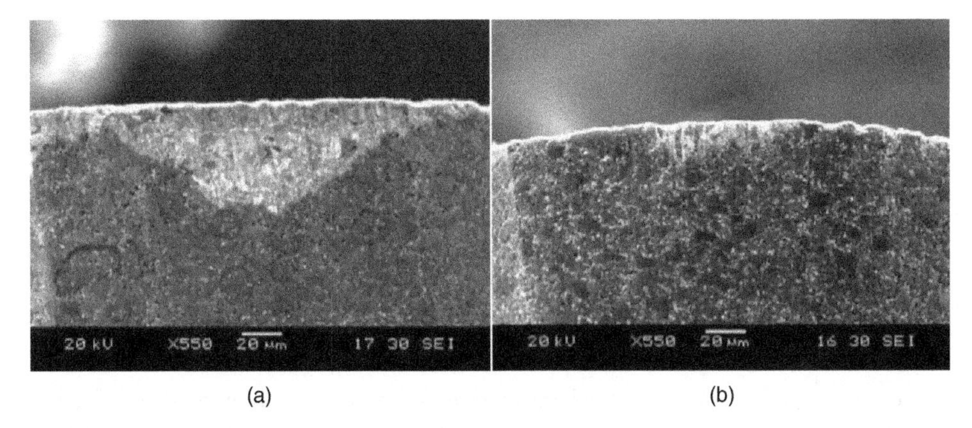

Figure 5.3 SEM photographs of worn-out tools: (a) conventional machining, (b) vibration-assisted machining. Source: Ming Zhou et al. [1]. © 2003, Taylor & Francis Group.

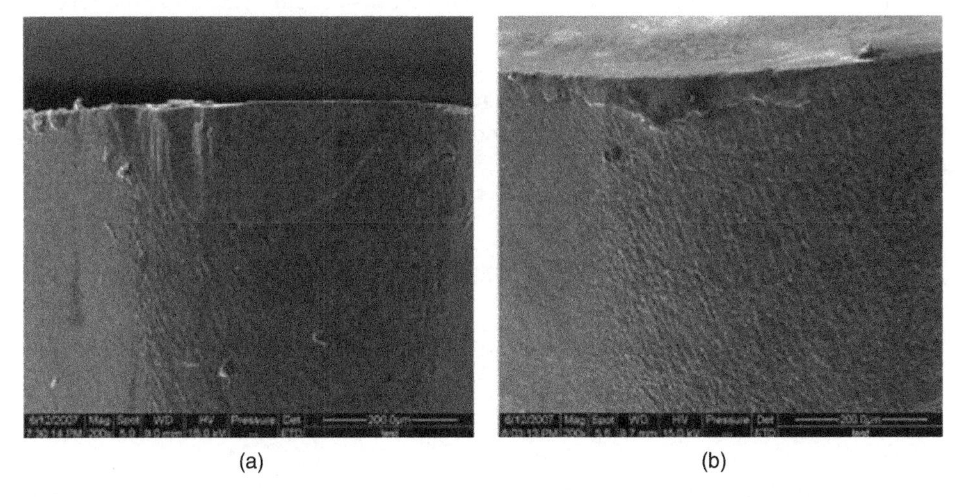

Figure 5.4 SEM micrograph of the tool flank wear patterns in (a) conventional turning, (b) vibration-assisted turning. Source: Dong et al. [2]. © 2013, Taylor & Francis Group.

and 5.4 show scanning electron microscope (SEM) images of wear comparison results for polycrystalline diamond (PCD) tools in conventional machining and 1D vibration-assisted machining [1, 2]. Scratches can be clearly seen, indicating abrasive wear as the dominant wear mechanism in these cutting experiments. Steel and SiCp/Al composites were selected as the workpiece materials in these cases. The cutting and vibration parameters of vibration-assisted machining in both experiments are set to satisfy the periodic separation conditions of the tool and the workpiece, which reduce the average cutting force and temperature. As a result, a dramatic reduction of tool wear area can be found for these experiments as vibration is added. Moreover, smoother wear can be found in the results of vibration-assisted machining while damage on the conventional machining tool is irregular. This also leads to better surface finish results for vibration-assisted machining.

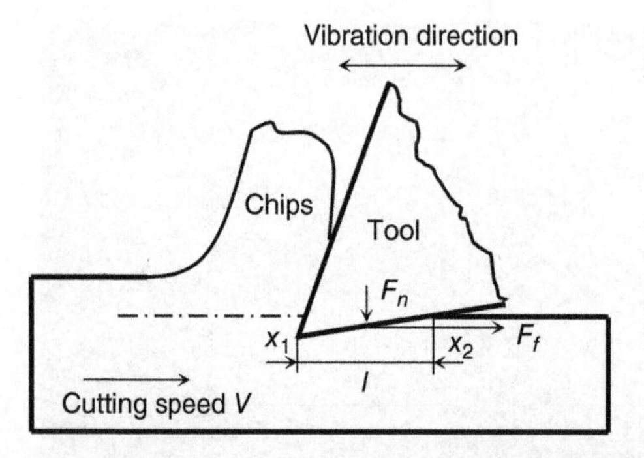

Figure 5.5 Relative motion between the workpiece and the tool in 1D vibration-assisted machining.

As discussed in Section 5.2.2, mechanical wear is mainly caused by the strong friction between the tool and workpiece and flank wear is often used as a standard for tool wear and tool life assessment. Therefore, analyzing the friction force between the tool flank face and workpiece can be significant for tool wear. The kinematics of cutting tool in vibration-assisted machining needs to be understood first. Figure 5.5 shows the relative motion between the workpiece and the tool in 1D vibration-assisted machining, and high-frequency reciprocating motion can be obtained by the cutting tool.

In Figure 5.5, the initial contact length between the tool flank face and workpiece is l, and when vibration is added on the cutting tool, the actual contact length can be expressed as

$$L = l + A \sin 2\pi ft \tag{5.1}$$

where t is the cutting time, and f and A are the vibration frequency and amplitude respectively.

Therefore, the relative position of the two contact points x_1 and x_2 can be expressed as

$$x_1 = Vt + \frac{L}{2} = Vt + \frac{l + A \sin 2\pi ft}{2} \tag{5.2}$$

$$x_2 = Vt - \frac{L}{2} = Vt - \frac{l + A \sin 2\pi ft}{2} \tag{5.3}$$

where V is the cutting speed.

The relative speed of the two contact points, V_1 and V_2, can be expressed as

$$V_1 = \dot{x}_1 = V + \pi fA \cos 2\pi ft = V + V_c \cos 2\pi ft \tag{5.4}$$

$$V_2 = \dot{x}_2 = V - \pi fA \cos 2\pi ft = V - V_c \cos 2\pi ft \tag{5.5}$$

where V_c is the critical velocity, which is defined in Eq. (3.4).

Then the total friction force on the tool flank face $F_t(t)$ can be expressed as

$$F_t(t) = \frac{\mu F_n}{2}[\text{sgn}(V + V_c \cos 2\pi ft) + \text{sgn}(V - V_c \cos 2\pi ft)] \tag{5.6}$$

where F_n is the fixed normal force between the two contact points, and μ is the coefficient of friction.

The average of the total friction force F_a during a vibration cycle can be expressed as

$$F_a = \frac{1}{T}\int_0^T F_t(t)dt = \begin{cases} \dfrac{2\mu F_n}{\pi}\arcsin\left(\dfrac{V}{V_c}\right) & (V < V_c) \\ \mu F_n & (V \geq V_c) \end{cases} \tag{5.7}$$

According to Eq. (5.7), the friction force on the tool flank face is μF_n when cutting speed V is larger than or equal to the critical velocity V_c. This process is similar to the normal machining process so the direction of the friction F_r does not change during the machining process. The friction force on the tool flank face is much less than μF_n when cutting speed V is less than the critical velocity V_c. The direction of friction changes periodically with the separation of the tool from the workpiece, which in turn reduces tool mechanical wear.

5.2.3.2 Mechanical Wear Suppression in 2D Vibration-Assisted Machining

Figure 5.6 shows the results of a comparison of tool wear in conventional machining and 2D vibration-assisted machining on Inconel 718 [3]. Scratches can be observed from the worn parts of the two cutting tools, which indicates that the main wear mechanism is abrasive wear. It can be found that the flank wear in conventional machining is considerably heavier than that in 2D vibration-assisted machining. In addition, the buildup of edge and boundary wear is effectively suppressed as 2D vibration is added. Similar results can also be found in Alfredo's research [4]. Tool life is significantly improved by adding elliptical vibrations to the face milling process of Ni-Alloy 718. Zhao et al. [5] studied the cutting mechanism of vibration-assisted machining on Inconel 718 by comparing the tool wear and chip morphology in conventional and 2D vibration-assisted machining at different cutting lengths. The results indicate that the effective reduction of abrasive wear is one of the reasons for the longer tool life in 2D vibration-assisted machining. Figure 5.7 shows a comparison of the wear area of the steel processed by diamond tools in conventional machining, 1D vibration-assisted machining, and 2D vibration-assisted machining [6]. Smaller tool wear can be observed when vibration is added. Slight tool flank wear exists around the tool edge in 1D vibration-assisted machining results, while the tool wear is almost undetectable in 2D vibration-assisted machining. In order to further study the tool wear mechanism, energy-dispersive X-ray spectroscopy (EDS) analysis is conducted on the worn area on the tool flank faces and the results indicate that the tool–workpiece separation in 2D vibration-assisted machining promotes the generation of the oxide layer on the freshly machined surface, which in turn reduces the mechanical wear effect of tool edge.

The above experimental results indicate that the mechanical wear of the cutting tool in 2D vibration-assisted machining can be further reduced compared with 1D vibration-assisted machining results. Besides the reasons given above, two merits, lower cutting force and a unique elliptical tool trajectory, also contribute to this phenomenon. Figure 5.8 shows cutting force variation in 1D vibration-assisted machining and 2D vibration-assisted machining respectively [7]. It can be found that the cutting force in 2D vibration-assisted machining is significantly reduced compared with that in 1D vibration-assisted machining. Other published results also indicate the positive effect on cutting force reduction of vibration-assisted machining [8–12]. On the other hand, the tool–workpiece contact area is not fixed during each elliptical cutting cycle. Figure 5.9 shows the relative motion between the workpiece and the tool in 2D vibration-assisted machining and three tool positions during machining

Figure 5.6 SEM micrographs of cutting tools used in (a) conventional machining and (b) elliptical vibration-assisted machining. Source: Lu et al. [3]. © 2015, Taylor & Francis Group.

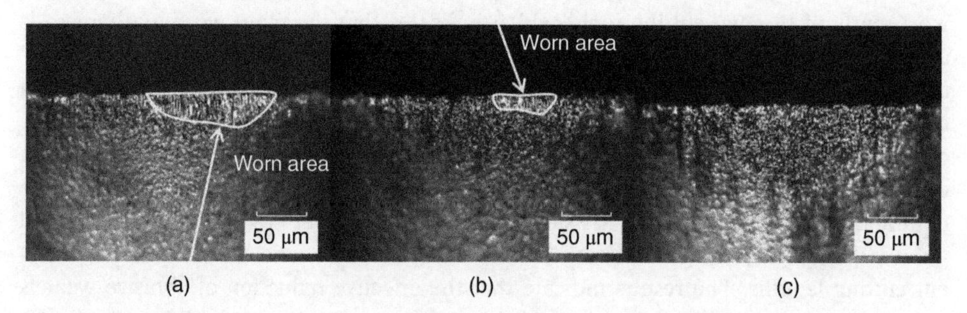

Figure 5.7 Microscope photographs (×1000) of the tool flank faces in the three machining processes: (a) conventional machining, (b) 1D vibration-assisted machining, and (c) 2D vibration-assisted machining. Source: Zhang et al. [6]. © 2014, Elsevier.

are selected to vary the tool–workpiece contact area. It can be found that the flank wear area of the cutting tool decreases in each elliptical cutting cycle. In addition, the rake angle of the cutting tool gradually becomes more negative with the tool moving upward in its elliptical trajectory, leading to an inconstant shear angle for the processing material [13, 14]. Moreover, cracks are generated in the cutting area of the hard and brittle workpiece due to the fluctuation in the cutting thickness and the impact of the cutting force. These are also reasons for the smaller cutting force in 2D vibration-assisted machining than in1D vibration-assisted machining. As a result, the uncut material is easier to remove and the tool life in 2D vibration-assisted machining is further extended.

5.2.3.3 Thermochemical Wear Suppression in Vibration-Assisted Machining

Thermochemical mechanisms appear to be the main cause of wear when diamond tools are used to cut nonferrous metals, especially for the hard and brittle materials. Taking titanium alloys as an example, the severe friction and elastic deformation during the processing of the titanium materials by diamond tools produce high cutting temperatures and contact stresses. On the other hand, the low thermal conductivity of titanium alloy leads to a large accumulation of cutting heat on the processing area, which not only causes the diamond

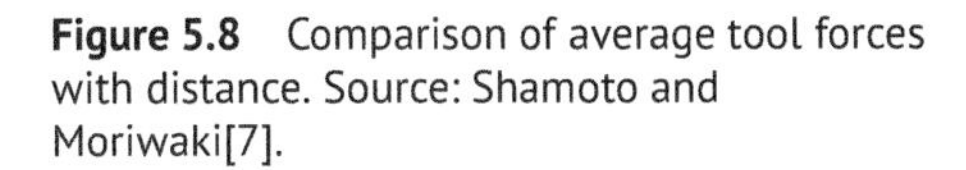

Figure 5.8 Comparison of average tool forces with distance. Source: Shamoto and Moriwaki[7].

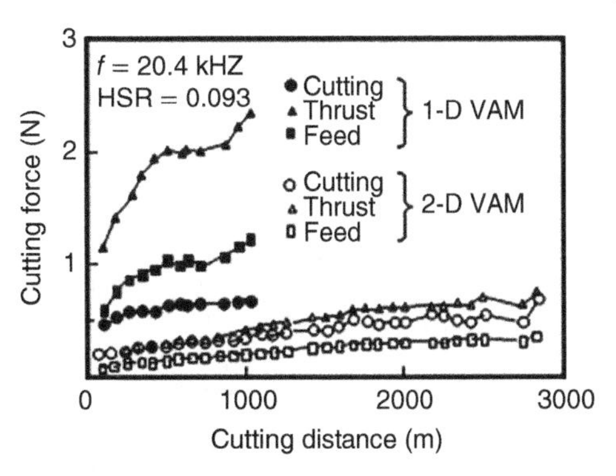

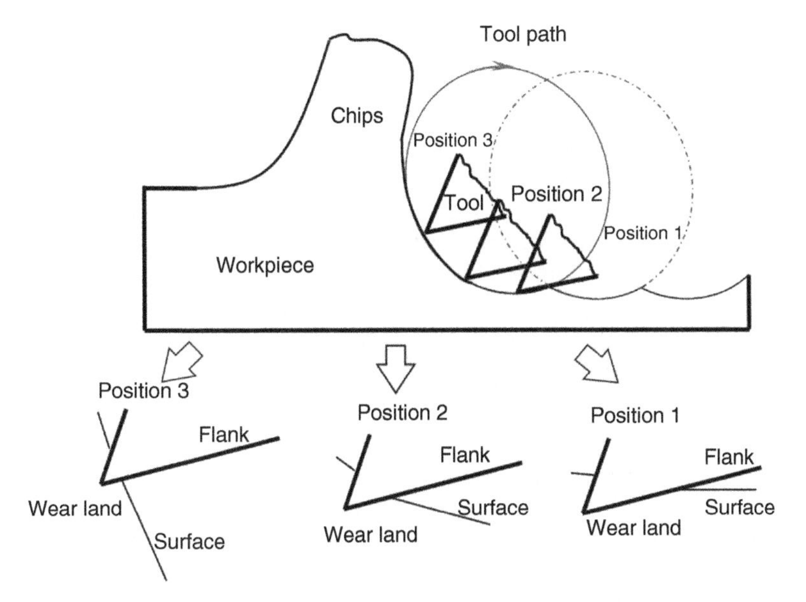

Figure 5.9 Relative motion between the workpiece and the tool in 2D vibration-assisted machining.

tool material to diffuse into the titanium alloy material but also causes it to chemically react with oxygen in the air. Figure 5.10 shows the schematic diagram of the contact area between the diamond tool and the titanium alloy workpiece during processing. Diffusion wear mainly occurs in zone I due to the continuous and close contact between the tool edge and the chips, which generates an environment of high temperature and pressure at the tool tip during the material processing, leading to the tool material easily reacting with the workpiece material. At the same time, oxygen can easily enter zone II and III, which exacerbates the oxidation wear in these zones. As discussed in Section 5.2.2, diffusion wear is caused by the mutual diffusion of particles between the cutting tool and the workpiece. According to Fick's first law, the concentration gradient is a prerequisite for the occurrence of the diffusion process and the diffusion rate is proportional to temperature. During the cutting process, the diamond tool remains in contact with the freshly machined surface

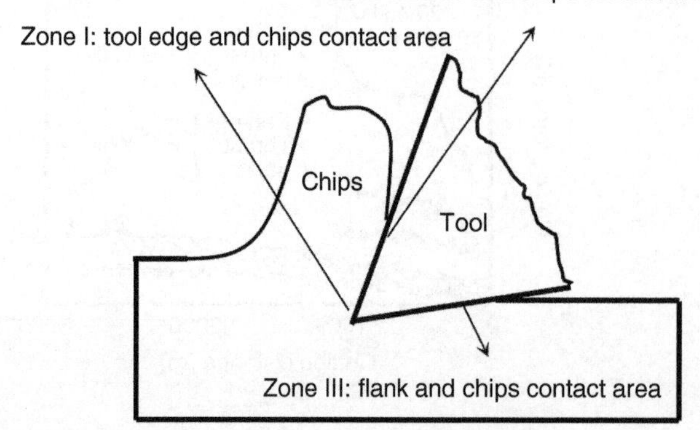

Figure 5.10 Schematic diagram of the contact area between the cutting tool and the workpiece.

of the workpiece, leading to the carbon atoms in the diamond tool diffusing into the titanium alloy workpiece due to the effect of cutting heat and the concentrate gradient. At the same time, the continuous flow of the workpiece material in the cutting deformation zone maintains a large diffusion flux between the diffusion interfaces, and the strong plastic deformation of the workpiece material increases the dislocation density and the voiding, which further increases the diffusion effect of the tool material. At a specific temperature, if the concentration of the diffusion element in the workpiece material exceeds its solubility, the tool material will chemically react with the workpiece material, and the diffusion reaction will be converted into a chemical reaction.

The thermochemical wear of diamond tools and the direction and strength of the chemical reaction involved are inseparably intertwined during the cutting process. As one of the calculation methods in thermodynamics, the Gibbs free energy function method simplifies the calculation of the chemical reaction and energy balance, which greatly reduces the computational workload and complexity. Therefore, the analysis of the standard Gibbs free energy of the chemical reaction that may occur during the cutting process at different temperatures can be used to determine the tendency toward thermochemical wear in the tool [15].

For a diffusion wear process, the free energy generated by its diamond tool material can be expressed as [16]

$$\Delta G_f^0 = \Delta G^{xs} + RT \ln c \tag{5.8}$$

where ΔG_f^0 is the free energy generated by diamond tool material (J/mol), ΔG^{xs} is the excess free energy of diamond tools in the titanium alloy workpiece (J/mol), R is the universal gas constant, and T is the temperature in Kelvin.

According to Eq. (5.8), the solubility of the diamond material in the titanium alloy is

$$c = e^{\frac{\Delta G_f^0 - \Delta G^{xs}}{RT}} \tag{5.9}$$

Ti–6Al–4V is an $(\alpha + \beta)$ titanium alloy in which almost 80% is α titanium. Therefore, in the cutting process, the solubility of diamond tool material (carbon element) in α titanium is mainly considered and its solubility is 1% at 920 °C [17]. In addition, diamond is a

Table 5.1 Gibbs free energy of carbon at different temperatures.

T (K)	298	500	700	900	1100	1300	1500
ΔG_f^0	−2.892	−3.577	−4.255	−4.933	−5.611	−6.289	−6.967

Source: Paul et al. [18].

Table 5.2 The solubility of carbon in titanium alloy under different temperature.

T (K)	298	500	700	900	1100	1300	1500
c	3.38E−08	2.99E−05	5.22E−04	2.60E−03	7.00E−03	1.42E−02	−2.37E−02

Source: Shao et al. [19].

metastable crystal structure of carbon and graphite is a stable crystal structure of carbon. The crystal structures of diamond will turn to the crystal structures of graphite when the carbon atoms are detached. The diffusion process of carbon atoms will occur with the diamond getting graphitized since the carbon atoms in the diamond are arranged more closely [18]. According to Gibbs–Helmholtz equations, the Gibbs generation free energy of carbon in the diamond tool, in other words, Gibbs free energy of diamond into graphite at different temperatures, is listed in Table 5.1.

The excess free energy of diamond tools (carbon element) in titanium alloy workpiece [19] $\Delta G^{xs} = 39.727$ kJ/mol. Therefore, the solubility of carbon in titanium alloy under different temperatures is shown in Table 5.2.

It can be found from Table 5.2 that the solubility of carbon in a titanium alloy will increase as temperature increases, which is in good agreement with Fick's first law. In other words, a higher cutting temperature can speed up the diffusion wear of the diamond tool in processing the titanium alloy. Thus, cutting tool life can be extended effectively by reducing the cutting temperature and the contact time between the cutting tool and workpiece, which is also one of the reasons why the cutting fluid can effectively reduce tool wear during the ultraprecision machining of titanium alloy.

In the vibration-assisted machining process, the contact time between the cutting tool and workpiece can be reduced due to the periodic separation between the tool and workpiece, which also allows sufficient cutting fluid or air to enter the cutting zone, improving lubrication conditions and leading to lower cutting temperatures. As a result, the reaction rates can be reduced and lower tool wear can be obtained.

As an important method of distinguishing between different components in a sample, energy-dispersive X-ray spectroscopy (EDS) is commonly used to identify the thermochemical wear of a worn tool by analyzing the elements on the surface of the tool. Figure 5.11 shows the tool wear comparison results for conventional and vibration-assisted milling. Figure 5.3c shows the EDX results of the worn tool surface and indicates that the material bonded to the tool is the workpiece material Ti–6Al–4V. Heavy bonded workpiece material can be found in the cutting tool of conventional machining, while much less bonded workpiece material is found in the vibration-assisted machining results, which indicates that the effect of adhesion wear on the tool is effectively reduced as the vibration is added. Similarly,

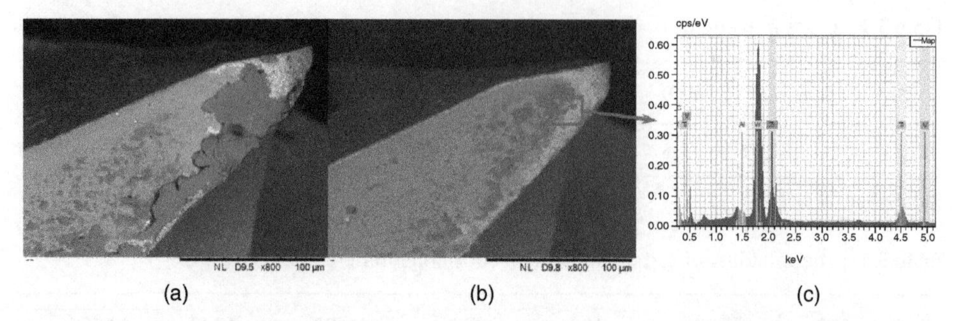

Figure 5.11 SEM micrographs of cutting tools: (a) conventional milling, (b) vibration-assisted milling, and (c) EDX results. Source: Courtesy of Dehong Huo, Wanqun Chen, and Lu Zheng.

other experimental results also indicated the positive effect of vibration-assisted machining in reducing thermochemical wear [20, 21].

5.2.3.4 Tool Wear Suppression in Vibration-Assisted Micromachining

Increasing demand for micro-parts and components has led to the development of micromachining technology. Micromachining involves higher technical requirements for the cutting tool, and tool life is one of the most important factors restricting its application. Owing to the well-known size effect, the cutting edge radius can be no longer ignored, and this leads to a different cutting mechanism in micromachining compared to its macroscale counterpart process. Machining performance can be influenced by the tool edge radius when the undeformed chip thickness is small enough. It has been reported that the ratio of undeformed chip thickness to cutting edge radius is closely linked to the ploughing effect on the machining surface, effective rake angle, and specific cutting energy, which in turn affects the cutting performance.

According to the relationship between the uncut chip thickness H and the minimum chip thickness Hc, which is largely determined by the machining parameters, the inherent material properties and cutter edge radius re, the circumstances of the micromachining process can be divided into three types as shown in Figure 5.12. When the uncut chip thickness is smaller than the minimum chip thickness, as shown in Figure 5.12a, no chips are formed and only plastic–elastic deformation occurs in the cutting area, which intensifies the ploughing effect and tool wear. Chips start to form when the value of uncut chip thickness reaches that of the minimum chip thickness, as shown in Figure 5.12b. When the uncut chip thickness is comparable to the minimum chip thickness, a mixed deformation type that combines plastic–elastic and shear deformation happens in the cutting zone. As the uncut chip thickness increases toward the minimum chip thickness, as shown in Figure 5.12c, the workpiece material is removed by the cutter as a chip, and the elastic recovery becomes very small and negligible. In this process, the surface finish quality, tool life, surface roughness, and burr formation could be improved by increasing the uncut chip thickness. According to the current research results, the minimum chip thickness H is usually 0.2–0.4 times the value of re [23–26].

The mechanical and thermochemical wear mechanisms are still effective in explaining the wear of micro-tools. Meanwhile, ploughing effect is the main reason for micro-tool wear when the feed rate is small enough. However, the effect of minimum chip thickness is

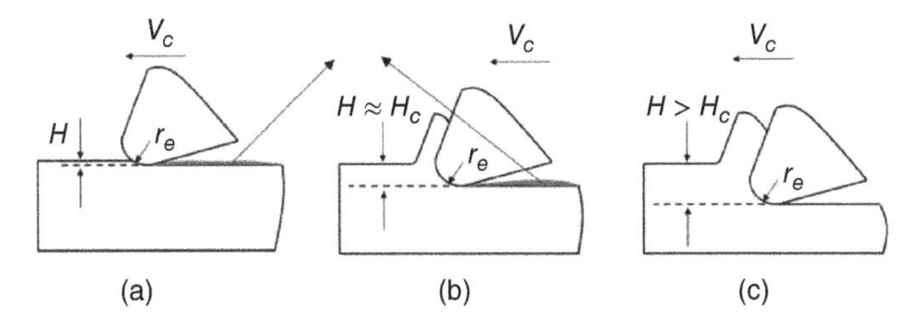

Figure 5.12 Material removal mechanism in micromachining process: (a) elastic deformation, (b) plastic–elastic deformation, and (c) plastic deformation. Source: Zheng et al. [22].

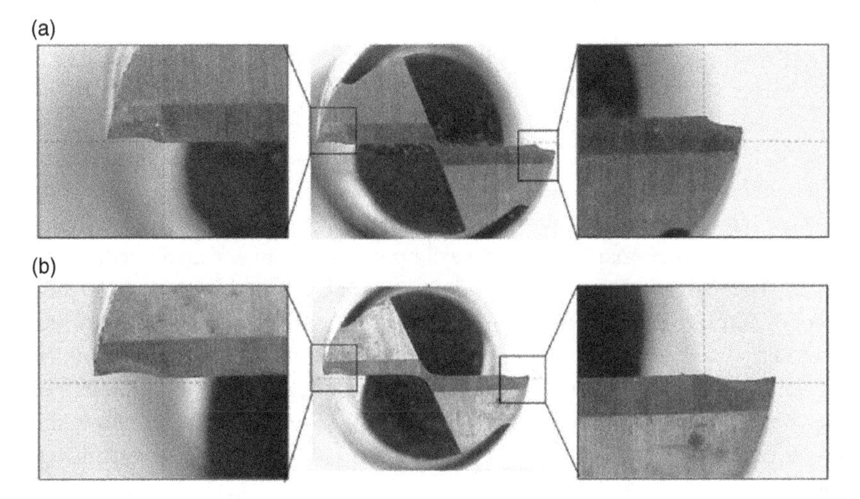

Figure 5.13 Photos of tool wear: (a) without vibrations and (b) with vibrations. Source: Ding et al. [27]. © 2010, Elsevier.

unavoidable in a micromachining process. By applying vibration to the micromachining process, its stability can be improved because the instantaneous uncut chip thickness and cutting speed are changed significantly compared with the conventional micromachining process, especially in the cutting in and cutting out area. As a result, the duration of the period of squeeze friction between the tool and the workpiece can be reduced, resulting in longer tool life. Figure 5.13 shows micro-tool wear comparison results between conventional micro-milling and vibration-assisted milling [27]. The results show that the tool wear with vibration assistance can be reduced by approximately 5–20% compared to that in conventional micro-milling. Moreover, the machined surface roughness and surface finish from vibration-assisted processing is also better than in the conventional micro-milling results, since the ploughing effect is effectively suppressed, which reduces the secondary damage from the worn cutting tool on the machined surface.

5.2.3.5 Effect of Vibration Parameters on Tool Wear

Vibration parameters, such as vibration direction, amplitude, and frequency have a great influence on tool life in vibration-assisted machining. Changes in the vibration parameters will change the tool path, resulting in different cutting mechanisms, which in turn

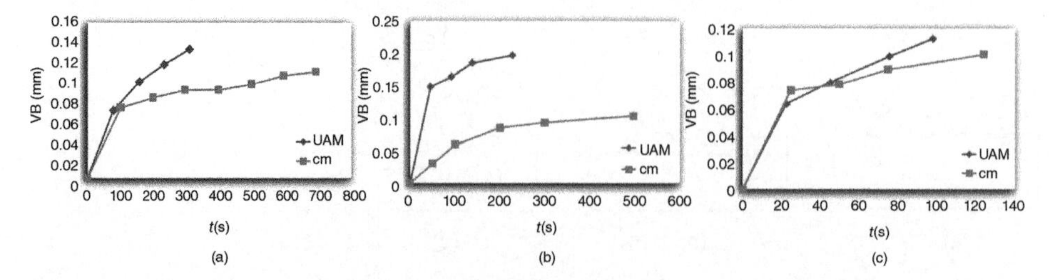

Figure 5.14 Time history of flank wear land width in conventional machining and vibration-assisted machining: (a) spindle speed 500 r/min, (b) spindle speed 1000 r/min, and (c) spindle speed 2000 r/min. Source: Janghorbanian et al. [28].

influence the tool life. The relationship between vibration parameters and tool trajectory is introduced in Chapter 4. Improperly set vibration parameters can shorten tool service life and reduce machined surface roughness and finish. Therefore, the optimization of vibration parameters in a vibration-assisted machining process is necessary. Besides this, the relationship between the vibration parameters and other machining factors, such as the machining parameters and the size and geometry of the cutting tool, also need to be taken into account. Taking vibration-assisted micromachining as an example, in order to effectively reduce the ploughing effect, elliptical vibration with high frequency and small amplitude as needed to meet the tool–workpiece separation conditions is usually applied. However, when the amplitude is too large, excessive fluctuations in cutting force will affect the tool, which increases tool wear. Janghorbanian et al. [28] experimentally studied the variation in tool life in elliptical vibration-assisted milling at different cutting speeds. Figure 5.14 shows the tool flank wear experiment results at low spindle speed ($n = 500$, 1000, and 2000 r/min). It can be found that tool wear is more serious when vibration is added. This is due to the unfavorable impacts of the workpiece on the tool flank surface and, subsequently, mechanical and wear mechanisms are the main factors leading to rapid tool failure in vibration-assisted machining process. Lower spindle speeds lead to more effective impacts between the workpiece and tool flank surface. As spindle speed increases, the contact of tool on the workpiece also increases in each vibration cycle, and cutting approaches a continuous form, and consequently, the number of impacts on the tool flank surface is reduced. Moreover, the same results can also be obtained when vibration frequency is increased. Similar results have also been found elsewhere [29–32].

5.3 Burr Formation

Similar to chip generation, Burr formation is a common phenomenon in the machining process and an important criterion in the evaluation of the machined surface. Burrs are caused by the plastic deformation of the material at the end of the cutting process and their generation is affected by many factors, including cutting parameters, tool geometry, and material properties [33]. The presence of burrs affects the product life cycle and machining accuracy, and also poses a risk to operators. Therefore, an extra deburring or edge finishing process is usually needed, which is costly and inefficient. Moreover, the deburring process may introduce unwanted residual stress and can even damage the parts.

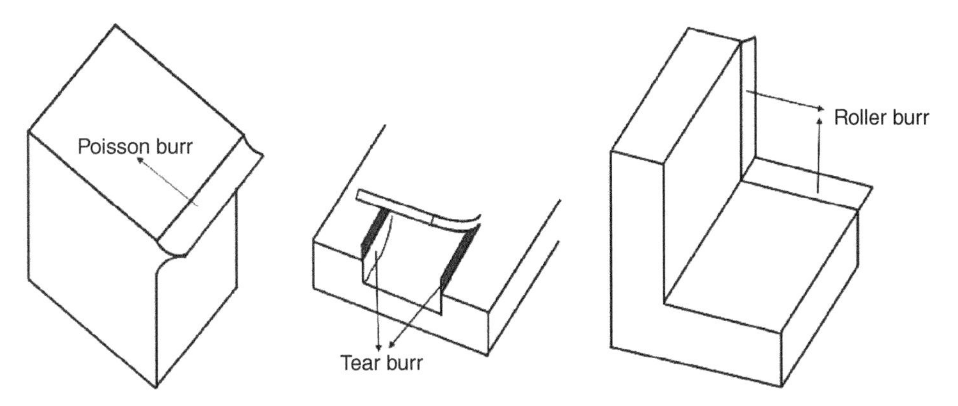

Figure 5.15 Schematics of Poisson burr, tear burr, and rollover burr. Source: Zheng et al. [35].

5.4 Burr Formation and Classification

Burr formation is a complex process influenced by many factors such as material properties, manufacturing processes, and shapes. The first study of burr formation in the machining process can be traced back to the 1970s [34]. It was found that the metal of the cutting layer will produce a large shear slip under the action of the tool, resulting in a large plastic deformation in the cutting process. A burr is formed if the material remains on the workpiece. Currently, two different approaches are commonly accepted by researchers for descriptions of burring in machining. In terms of burr formation, four types of machining burrs can be categorized: rollover burr, tear burr, cut-off burr, and Poisson burr, as shown in Figure 5.15. The rollover burr is a kind of bended chip, which is also called an exit burr because it is usually formed at the end of processing. The tear burr is not the result of shearing clearing; rather it is caused by the result of material tearing loose from the workpiece, which is much like the burr formation found in the punching process. The cut-off burr is the consequence of workpiece falling apart from the raw material before the cutting process is finished. The Poisson burr is a result of the material's tendency to bulge to the sides when it is compressed until permanent plastic deformation occurs. Another type of burr description is according to the shapes, locations, and formation mechanisms of the burrs [36]. Figure 5.16 shows the burr types in the machining process, which includes entrance burr, exit burr, side burr, and top burr. The size and formation mechanisms of these four types of burrs are different. The top burr is the result of Poisson and tear burr and appears at the top surface of the workpiece. The exit burr is defined as a burr attached to machined edge at the end of milling and the side burr adheres to the transition surface [37].

5.5 Burr Reduction in Vibration Assisted Machining

Conventional machining is a continuous cutting process and the formation of chips and burrs is the process of the deformation of the workpiece material in the cutting layer. Figure 5.17 shows the layout of four workpiece material deformation zones. The first three zones are directly related to the quality of the machined surface and chip formation.

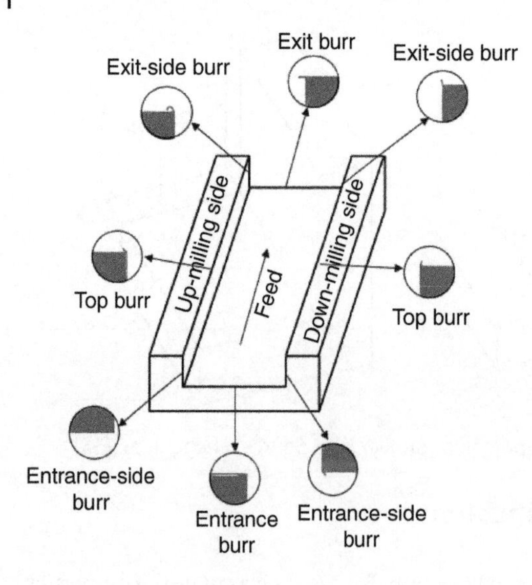

Figure 5.16 Types of burrs in micro-end milling. Source: Zheng et al. [35].

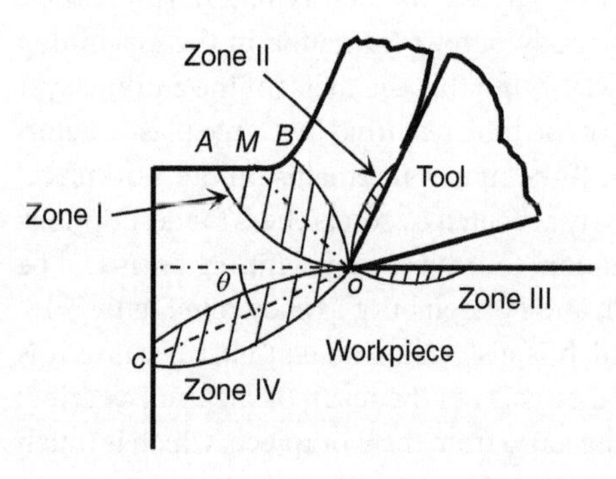

Figure 5.17 Layout of workpiece material deformation zone.

In zone I, also called primary shear zone, workpiece material has been deformed by a concentrated shearing process and chip formation takes place as the cutting process continues. The chip and the rake face of the tool are in contact in zone II, also known as the secondary shear zone. The workpiece material starts to flow when the frictional stress on the rake face is equal to the shear yield stress of the workpiece material. In zone III, also called tertiary shear zone, the clearance face of the tool will rub against the newly machined surface and cause its deformation.

As the cutting tool approaches the end of the workpiece, a downwardly extending deformation zone is generated due to the reduced support stiffness in that area. This area is zone IV, also called negative shear band, which has an important influence on the burr generation. The negative shear band extends to point c at the end of the workpiece and this point is called the plastic hinge point. The position of the plastic hinge point c has a significant influence on burr formation and the workpiece material below the plastic hinge no longer undergoes plastic deformation. In addition, the negative shear angle θ can be obtained due to the shear and slip effect and elastic–plastic deformation in this zone. A higher value of the negative shear angle θ and the greater deformation of the end of the workpiece in turn lead to the formation of larger burrs.

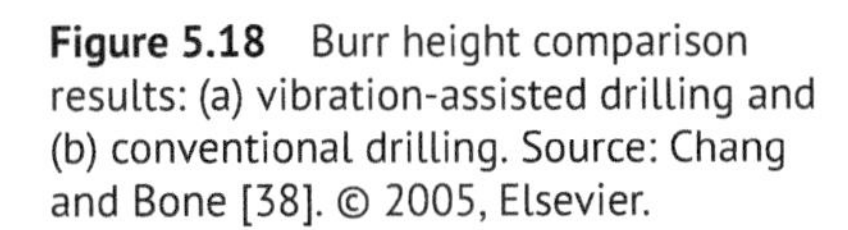

Figure 5.18 Burr height comparison results: (a) vibration-assisted drilling and (b) conventional drilling. Source: Chang and Bone [38]. © 2005, Elsevier.

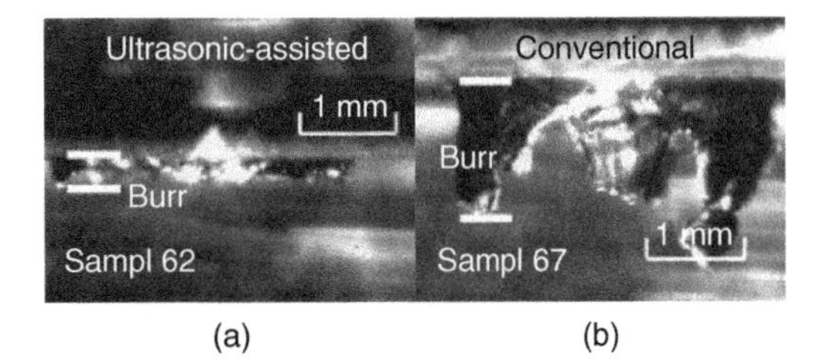

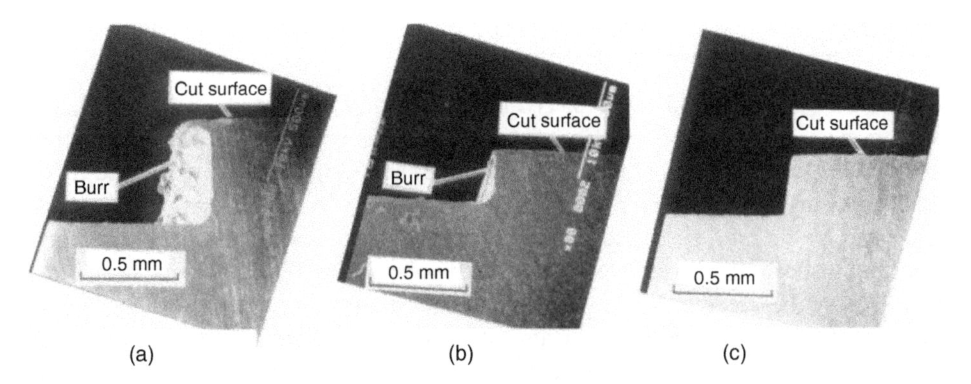

Figure 5.19 SEM photographs of burrs on the workpiece edges: (a) conventional cutting; (b) 1D vibration cutting; (c) elliptical vibration cutting. Source: Ma et al. [39]. © 2005, Elsevier.

Figure 5.18 shows the burr height comparison results between conventional and vibration-assisted drilling [38]. It can be found that burr formation can be reduced dramatically in vibration-assisted machining process. Noncontinuous cutting can be achieved in vibration-assisted machining due to the rapid variation on the cutting speed, leading to a pulse cutting process. The probability and time of the bending and tearing of the workpiece material can be effectively reduced due to the higher cutting energy. As a result, small pieces of chips can be generated and burr formation can be effectively suppressed. Moreover, because of the unique tool trajectory, burr formation in 2D vibration-assisted machining can be further suppressed. Figure 5.19 shows the side burr formation in turning processes for conventional cutting, 1D vibration-assisted machining, and 2D vibration-assisted machining, and the smallest burr can be found in the 2D vibration-assisted machining results [39]. This is due to the lower instantaneous compressive and bending stresses in the deformation zone in 2D vibration-assisted machining. In addition, the position of plastic hinge point *c* will move toward the machined surface in each elliptical cutting cycle, reducing the elastic–plastic deformation area at the end of the workpiece, which in turn also reduces burr formation.

5.5.1 Burr Reduction in Vibration-Assisted Micromachining

The removal of burrs is more difficult in micromachining due to the small size of the workpiece. As discussed above, the size effect in micromachining leads to a different cutting mechanism, and this influences burr formation. Figure 5.20 shows the formation of the machined surface in micromachining. In the micro-cutting process, compression shear

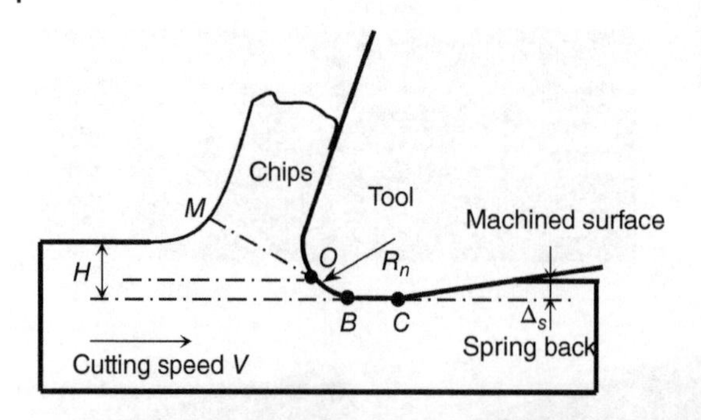

Figure 5.20 The formation of machined surface in micromachining.

deformation occurs when the workpiece material is compressed by the cutting tool, and chips are formed along the shear plane *OM* due to the shear slip effect. However, a layer of workpiece material Δs cannot form chips along the shear plane *OM* due to the size effect, and the machined surface can be obtained as the workpiece material is severely squeezed and rubbed by the cutting tool. As the micro-cutting tool approaches the end of the workpiece, the plastically deformed layer of the workpiece material Δs will flow and form burrs at the edges of the workpiece. Burr size will increase with the increase in the ratio of undeformed chip thickness to the cutting edge radius [40]. In addition, the point *O* will move up for a worn cutting tool due to the increase in cutting edge radius, which increases the thickness of Δs and the corresponding elastic–plastic deformation. As a result, the burr size will also be increased.

The size effect can be effectively reduced in vibration-assisted machining, leading to a smaller size of the burrs. Figure 5.21 shows a comparison between typical machined slots by 1D conventional and vibration-assisted micro-milling respectively [35]. Large top burrs can be found on the workpiece machined by conventional micro-milling on both the up-milling and down-milling sides. More machining tearing burrs appear on the down-milling side compared with the up-milling side and this phenomenon can be explained through an analysis of the slot milling process. As the cutter engages with the workpiece, the workpiece material is squeezed and pushed at the up-milling side first. As the process goes on, the uncut chip thickness increases, which enhances support effect on the uncut materials and the shear action on material removal. However, on the down-milling side, the support effect of the uncut material is smaller than that on the up-milling side, and hence the uncut materials are pushed out of the top of the down-milling side and large irregular tearing burrs are generated on this side. Meanwhile in 1D vibration-assisted micro-milling, no large top burr can be found on either side. Since vibration is superimposed on the process, the motion of tool tip becomes reciprocating, leading to a periodic separation between tool tip and the workpiece. As a result, the down-milling and up-milling can occur on both sides of the slot alternately [41]. In addition, the material removal mode changes from a shear dominated deformation in conventional micro-milling process to a mixed action that combines dynamic impact and shear deformation in vibration-assisted machining, and the ploughing/rubbing between workpiece and cutter is also reduced. As a consequence, the cutting force and discontinued chips can be reduced, which in turn reduces the burr generation and improves machining accuracy.

Figure 5.22 shows enlarged images on the down-milling side for tear burr observation between 1D vibration-assisted micromachining and 2D vibration-assisted machining [35].

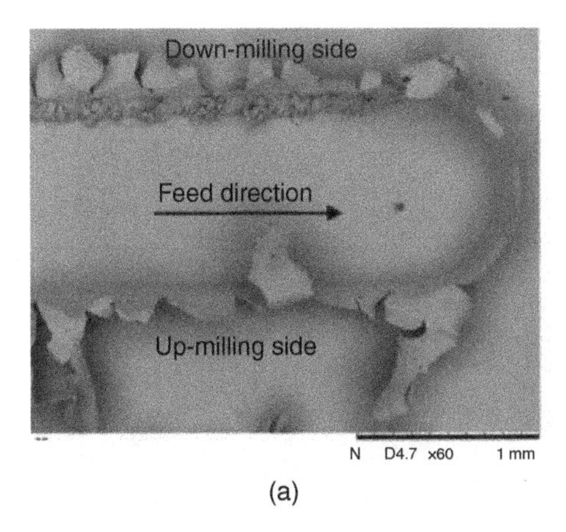

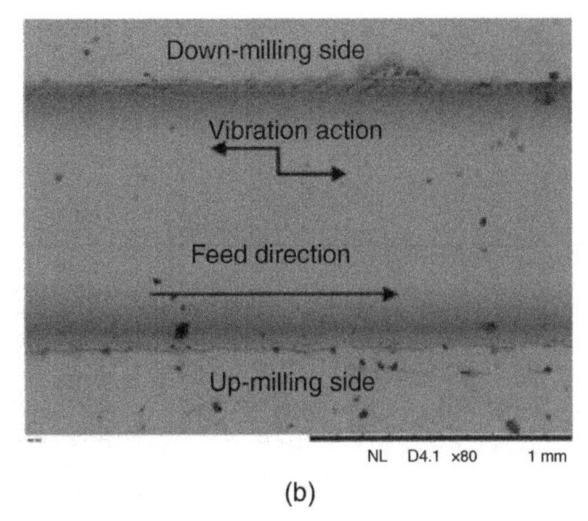

Figure 5.21 Machining results comparison between (a) conventional micro-end milling and (b) vibration-assisted micro-end milling. Source: Zheng et al. [35]. © 2018, SAGE Publications, Inc.

It can be seen that when the vibration frequency is constant, the tear burr size is reduced by adding cross-feed direction vibration to the process; this is because the tool tip cutting direction has been changed slightly and the material tearing loose from the workpiece is enhanced. The chip thickness therefore quickly reaches and exceeds the minimum cutting thickness when the cutting tool cuts in and out of the workpiece, reducing the effect of squeeze friction between the tool and the workpiece. This in turn reduces the size effect [41]. In addition, a trend of reduced tear burr can be found when increasing vibration frequency in the same vibration direction. As the tool vibration frequency increases from 400 to 8000 Hz, the tool vibration speed is enhanced, leading to an increase in reciprocating cutting on the down-milling side and the large tear burr is removed. As a result, the tear burr size can be reduced.

5.6 Concluding Remarks

5.6.1 Tool Wear

This chapter reviews tool wear occurrence and mechanisms, as well as the tool wear suppression mechanisms in the vibration-assisted machining process. The thermochemical and mechanical wear of the tool can be effectively reduced in vibration-assisted machining, leading to an extended tool life. The main reasons leading to this phenomenon are as follows:

(1) The processing time is reduced due to the periodic separation between the cutting tool and workpiece, which enhances the cooling effect of the tool and extends the tool life.
(2) The size effect can be effectively reduced in vibration-assisted micromachining, which alleviates the severe friction between the tool and the workpiece and reduces the tool wear.
(3) In vibration-assisted machining, cracks are generated in the cutting area of the hard and brittle workpiece due to fluctuations in the cutting thickness and the impact of the cutting force, leading to lower cutting forces and reduced tool wear.

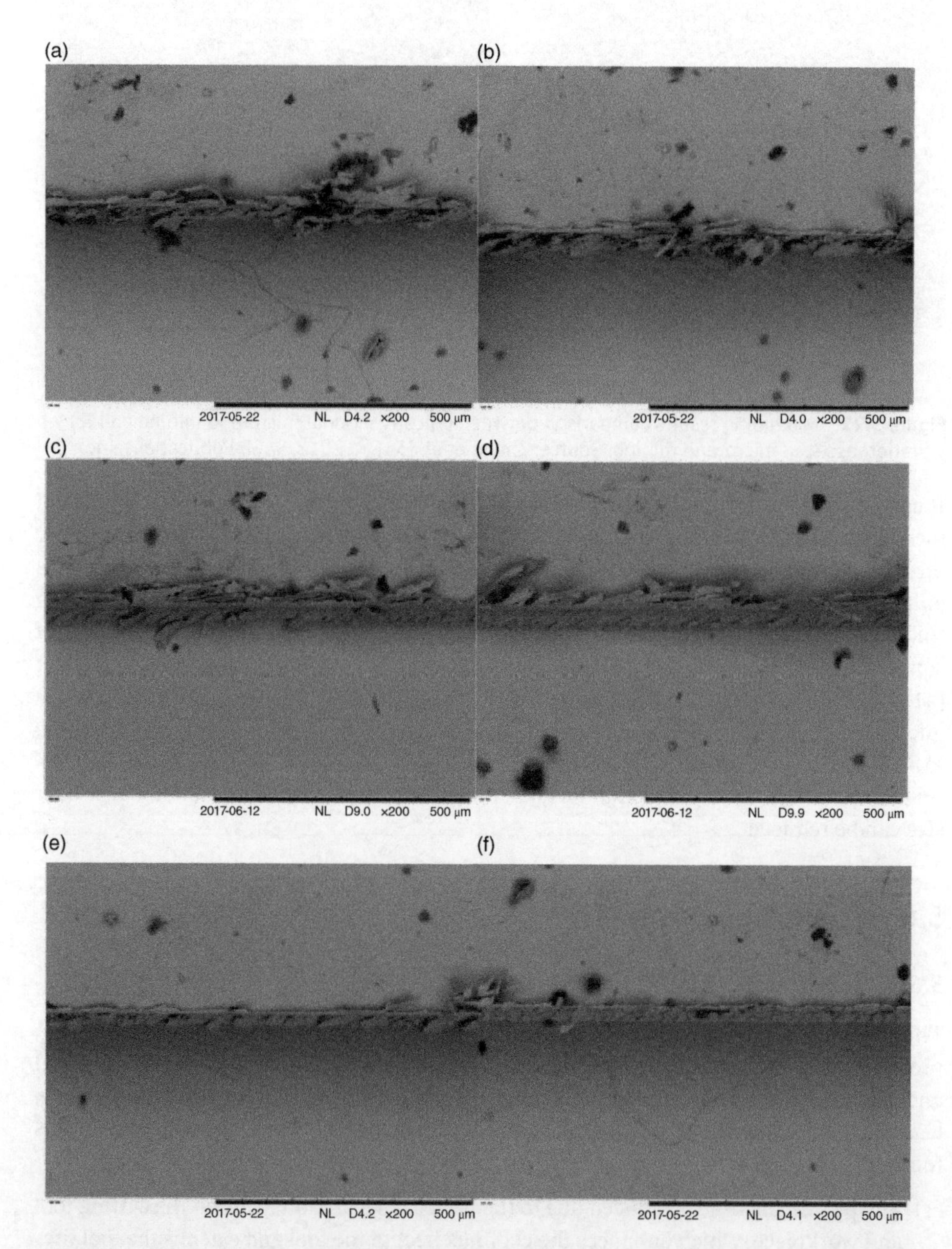

Figure 5.22 Down-milling side tear burr results with different parameters: (a) Feed direction vibration with a frequency of 400 Hz. (b) Feed and cross-direction vibration with a frequency of 400 Hz. (c) Feed direction vibration with a frequency of 4000 Hz. (d) Feed and cross-direction vibration with a frequency of 4000 Hz. (e) Feed direction vibration with a frequency of 8000 Hz. (f) Feed and cross-direction vibration with a frequency of 8000 Hz. Source: Zheng et al. [35]. © 2018, SAGE Publications, Inc.

5.6.2 Burr Formation

This chapter also reviews burr formation mechanisms, and the current development of burr formation suppression through vibration-assisted machining. The main reasons leading to this phenomenon are as follows:

(1) A lower friction and cutting temperature can be obtained due to the periodic separation between the cutting tool and workpiece in the vibration-assisted machining process, making the workpiece material easier to remove.
(2) Vibration-assisted machining is a noncontinuous cutting process, leading to rapid variations in cutting speed and the pulse cutting process. The probability and time of bending and tearing of the workpiece material can be effectively reduced due to the higher cutting energy. As a result, small pieces of chips can be generated and burrs formation can be effectively suppressed.
(3) In vibration-assisted micromachining, the chip thickness quickly reaches and exceeds the minimum cutting thickness when the cutting tool cuts in and out of the workpiece, reducing the effect of squeeze friction between the tool and the workpiece. This in turn reduces the size effect.

References

1 Zhou, M., Eow, Y.T., Ngoi, B.K.A., and Lim, E.N. (2003). Vibration-assisted precision machining of steel with PCD tools. *Mater. Manuf. Processes* 18: 825–834. https://doi.org/10.1081/AMP-120024978.

2 Dong, G., Zhang, H., Zhou, M., and Zhang, Y. (2013). Experimental investigation on ultrasonic vibration-assisted turning of SiCp/Al composites. *Mater. Manuf. Processes* 28: 999–1002. https://doi.org/10.1080/10426914.2012.709338.

3 Lu, D., Wang, Q., Wu, Y. et al. (2015). Fundamental turning characteristics of inconel 718 by applying ultrasonic elliptical vibration on the base plane. *Mater. Manuf. Processes* https://doi.org/10.1080/10426914.2014.973588.

4 Suárez, A., Veiga, F., de Lacalle, L.N.L. et al. (2016). Effects of ultrasonics-assisted face milling on surface integrity and fatigue life of Ni-alloy 718. *J. Mater. Eng. Perform.* https://doi.org/10.1007/s11665-016-2343-6.

5 Haidong, Z., Ping, Z., Wenbin, M., and Zhongming, Z. (2016). A study on ultrasonic elliptical vibration cutting of inconel 718. *Shock Vib.* https://doi.org/10.1155/2016/3638574.

6 Zhang, X., Liu, K., Kumar, A.S., and Rahman, M. (2014). A study of the diamond tool wear suppression mechanism in vibration-assisted machining of steel. *J. Mater. Process. Technol.* https://doi.org/10.1016/j.jmatprotec.2013.10.002.

7 Shamoto, E. and Moriwaki, T. (1999). Ultraprecision diamond cutting of hardened steel by applying elliptical vibration cutting. *CIRP Ann. – Manuf. Technol.* 48: 441–444. https://doi.org/10.1016/S0007-8506(07)63222-3.

8 Kong, X., Dong, J., and Cohen, P.H. (2017). Modeling of the dynamic machining force of vibration-assisted nanomachining process. *J. Manuf. Processes* 28: 101–108. https://doi.org/10.1016/j.jmapro.2017.05.028.

9 Kong, X., Zhang, L., Dong, J., Cohen, P.H. (2015). Machining force modeling of vibration-assisted nano-machining process. *ASME 2015 International Manufacturing Science and Engineering Conference*, MSEC 2015. doi:https://doi.org/10.1115/MSEC20159423.

10 Zhang, X., Senthil Kumar, A., and Rahman, M. (2012). Effects of cutting and vibration parameters on transient cutting force in elliptical vibration cutting. In: *Communications in Computer and Information Science*, 483–490. Springer https://doi.org/10.1007/978-3-642-35197-6_54.

11 Ni, C., Zhu, L., Ning, J. et al. (2019). Research on the characteristics of cutting force signal and chip in ultrasonic vibration-assisted milling of titanium alloys. *Jixie Gongcheng Xuebao/J. Mech. Eng.* 55: 207–216. https://doi.org/10.3901/JME.2019.07.207.

12 Qin, N., Pei, Z.J., Guo, D.M. (2009). Ultrasonic-vibration-assisted grinding of titanium: cutting force modeling with design of experiments. *Proceedings of the ASME 2009 International Manufacturing Science and Engineering Conference*, MSEC2009, pp. 619–624. doi:https://doi.org/10.1115/MSEC2009-84325.

13 Dautzenberg, J.H., Hijink, J.A.W., and van der Wolf, A.C.H. (1982). The minimum energy principle applied to the cutting process of various workpiece materials and tool rake angles. *CIRP Ann. – Manuf. Technol.* https://doi.org/10.1016/S0007-8506(07)63275-2.

14 Nakayama, K., Arai, M., and Kanda, T. (1988). Machining characteristics of hard materials. *CIRP Ann. – Manuf. Technol.* https://doi.org/10.1016/S0007-8506(07)61592-3.

15 Taminiau, D.A. and Dautzenberg, J.H. (2001). How to understand friction and Wear in mechanical working processes. *Int. J. Form. Processes* 4: 9–22.

16 Wong, T., Kim, W., and Kwon, P. (2004). Experimental support for a model-based prediction of tool wear. *Wear* 257: 790–798. https://doi.org/10.1016/j.wear.2004.03.010.

17 Hashmi, S. (2014). *Comprehensive Materials Processing*. Newnes https://doi.org/10.1016/c2009-1-63473-0.

18 Paul, E., Evans, C.J., Mangamelli, A. et al. (1996). Chemical aspects of tool wear in single point diamond turning. *Precis. Eng.* 18: 4–19. https://doi.org/10.1016/0141-6359(95)00019-4.

19 Shao, F., Liu, Z., and Wan, Y. (2010). Thermodynamical matching of alumina-based composite ceramic tools with typical workpiece materials. *Int. J. Adv. Manuf. Technol.* 49: 567–578. https://doi.org/10.1007/s00170-009-2413-0.

20 Zou, L., Huang, Y., Zhou, M., and Duan, L. (2017). Investigation on diamond tool wear in ultrasonic vibration-assisted turning die steels. *Mater. Manuf. Processes* 32: 1505–1511. https://doi.org/10.1080/10426914.2017.1291958.

21 Zhang, X., Huang, R., Liu, K. et al. (2018). Suppression of diamond tool wear in machining of tungsten carbide by combining ultrasonic vibration and electrochemical processing. *Ceram. Int.* 44: 4142–4153. https://doi.org/10.1016/j.ceramint.2017.11.215.

22 Zheng, L., Chen, W., and Huo, D. (2019). Experimental investigation on burr formation in vibration-assisted micro-milling of Ti-6Al-4V. *Proc. Inst. Mech. Eng. Part C J. Mech. Eng. Sci.* https://doi.org/10.1177/0954406218792360.

23 Malekian, M., Mostofa, M.G., Park, S.S., and Jun, M.B.G. (2012). Modeling of minimum uncut chip thickness in micro machining of aluminum. *J. Mater. Process. Technol.* https://doi.org/10.1016/j.jmatprotec.2011.05.022.

24 Kang, I.S., Kim, J.S., and Seo, Y.W. (2011). Investigation of cutting force behaviour considering the effect of cutting edge radius in the micro-scale milling of AISI 1045 steel. *Proc. Inst. Mech. Eng. Part B J. Eng. Manuf.* https://doi.org/10.1243/09544054JEM1762.

25 Vogler, M.P., DeVor, R.E., and Kapoor, S.G. (2004). On the modeling and analysis of machining performance in micro-endmilling, part I: surface generation. *J. Manuf. Sci. Eng. Trans. ASME* https://doi.org/10.1115/1.1813470.

26 Lai, X., Li, H., Li, C. et al. (2008). Modelling and analysis of micro scale milling considering size effect, micro cutter edge radius and minimum chip thickness. *Int. J. Mach. Tools Manuf.* https://doi.org/10.1016/j.ijmachtools.2007.08.011.

27 Ding, H., Ibrahim, R., Cheng, K., and Chen, S.J. (2010). Experimental study on machinability improvement of hardened tool steel using two dimensional vibration-assisted micro-end-milling. *Int. J. Mach. Tools Manuf.* 50: 1115–1118. https://doi.org/10.1016/j.ijmachtools.2010.08.010.

28 Janghorbanian, J., Razfar, M.R., and Zarchi, M.M.A. (2013). Effect of cutting speed on tool life in ultrasonic-assisted milling process. *Proc. Inst. Mech. Eng. Part B J. Eng. Manuf.* 227: 1157–1164. https://doi.org/10.1177/0954405413483722.

29 Pecat, O. and Brinksmeier, E. (2014). Tool wear analyses in low frequency vibration assisted drilling of CFRP/Ti6Al4V stack material. *Procedia CIRP* 14: 142–147. https://doi.org/10.1016/j.procir.2014.03.050.

30 Li, K.M. and Wang, S.L. (2014). Effect of tool wear in ultrasonic vibration-assisted micro-milling. *Proc. Inst. Mech. Eng. Part B J. Eng. Manuf.* 228: 847–855. https://doi.org/10.1177/0954405413510514.

31 Nath, C. and Rahman, M. (2008). Effect of machining parameters in ultrasonic vibration cutting. *Int. J. Mach. Tools Manuf.* 48: 965–974. https://doi.org/10.1016/j.ijmachtools.2008.01.013.

32 Zhang, X., Nath, C., Kumar, A.S. et al. (2010). A study on ultrasonic elliptical vibration cutting of hardened steel using PCD tools. In: *ASME 2010 International Manufacturing Science and Engineering Conference*, 163–169. MSEC https://doi.org/10.1115/MSEC2010-34239.

33 Ton, T.P., Park, H.Y., and Ko, S.L. (2011). Experimental analysis of deburring process on inclined exit surface by new deburring tool. *CIRP Ann. – Manuf. Technol.* 60: 129–132. https://doi.org/10.1016/j.cirp.2011.03.124.

34 Gillespie, L.K. and Blotter, P.T. (1976). The formation and properties of machining burrs. *J. Eng. Ind.* 98: 66. https://doi.org/10.1115/1.3438875.

35 Zheng, L., Chen, W., and Huo, D. (2018). Experimental investigation on burr formation in vibration-assisted micro-milling of Ti-6Al-4V. *Proc. Inst. Mech. Eng. Part C J. Mech. Eng. Sci.* https://doi.org/10.1177/0954406218792360.

36 da Silva, L.C., da Mota, P.R., da Silva, M.B. et al. (2015). Study of burr behavior in face milling of PH 13-8 Mo stainless steel. *CIRP J. Manuf. Sci. Technol.* 8: 34–42. https://doi.org/10.1016/j.cirpj.2014.10.003.

37 Piquard, R., D'Acunto, A., Laheurte, P., and Dudzinski, D. (2014). Micro-end milling of NiTi biomedical alloys, burr formation and phase transformation. *Precis. Eng.* 38: 356–364. https://doi.org/10.1016/j.precisioneng.2013.11.006.

38 Chang, S.S.F. and Bone, G.M. (2005). Burr size reduction in drilling by ultrasonic assistance. *Rob. Comput. Integr. Manuf.* https://doi.org/10.1016/j.rcim.2004.11.005.

39 Ma, C., Shamoto, E., Moriwaki, T. et al. (2005). Suppression of burrs in turning with ultrasonic elliptical vibration cutting. *Int. J. Mach. Tools Manuf.* 45: 1295–1300. https://doi.org/10.1016/j.ijmachtools.2005.01.011.

40 Aramcharoen, A. and Mativenga, P.T. (2009). Size effect and tool geometry in micromilling of tool steel. *Precis. Eng.* 33: 402–407. https://doi.org/10.1016/j.precisioneng.2008.11.002.

41 Chen, W., Teng, X., Zheng, L. et al. (2018). Burr reduction mechanism in vibration-assisted micro milling. *Manuf. Lett.* 16: 6–9. https://doi.org/10.1016/j.mfglet.2018.02.015.

6

Modeling of Cutting Force in Vibration-Assisted Machining

6.1 Introduction

Vibration-assisted machining (VAM) has the advantage of reducing the cutting forces involved. Analysis of the cutting force of VAM is useful in characterizing the cutting process, since tool wear and surface texture levels depend closely on the cutting force [1]. However, it is difficult to predict the transient cutting force of VAM due to its complicated motion trajectory. The study of chip formation is helpful in predicting the cutting force. The geometric features of chips have important effects on cutting force, and in order to investigate the time-varying cutting force in VAM, a model of the geometric features of chips is established based on an analysis of chip formation. The effects of cutting parameters on the geometric features of chips are then analyzed. To predict the transient force quickly and effectively, the geometric features of chips are introduced into the cutting force model.

Much research has been devoted to determining the cutting forces involved in VAM. Brehl and Dow [2] considered that the sectional area of the chip and properties of the workpiece material affect the cutting force, for which they established a model by studying variations in the contact area between the tool and the workpiece. However, Zhang et al. [3, 4] considered that the cutting force could be determined by combining the geometric features of chips and the tool, and the volume of chips removed was used as an indicator to predict the cutting force. Zhang et al. [5] carried out a cutting experiment on hardened steel using ultrasonic elliptical vibration cutting (UEVC) technology, and analyzed the effects of nominal cutting depth, feed rate, and cutting speed on cutting force, tool wear, and workpiece surface integrity. The experimental results showed that the cutting speed had the greatest effect on the cutting process. An analytical expression of the chip thickness of micro-end milling was first derived by Bao and Tansel [6]. Their model was widely used for the modeling of micro-milling cutting force. However, given the complex and intermittent cutting process in vibration-assisted milling (VAMILL), the values of amplitude and frequency of cutting force are different from those in conventional milling. Very little research on cutting force modeling for VAMILL can be found in the literature. Ding et al. [1] proposed a three-dimensional cutting force model for two-dimensional vibration-assisted micro-end milling, which considers the effects of the relative trajectory of the tool tip in relation to the workpiece. Zarchi et al. [7] investigated the effect of cutting speed and vibration amplitude on cutting forces in ultrasonic-assisted milling, and pointed out that as the cutting speed increases, the effect

Vibration Assisted Machining: Theory, Modelling and Applications,
First Edition. Lu Zheng, Wanqun Chen, and Dehong Huo.

of ultrasonic vibration on milling process tends to be insignificant, while the cutting forces in ultrasonic-assisted milling and conventional milling processes are similar to each other. Shen et al. [8] investigated the instantaneous uncut chip thickness in ultrasonic-assisted milling when ultrasonic vibration was applied in the feed direction, but only one previous cutting path was considered in their model.

Unlike the research methods mentioned above, the present study proposes a method using the geometric features of chips to predict the time-varying cutting force quickly and effectively. The aim is to establish a model of the geometric features of chips by analyzing chip formation, and then to find the relationship between the geometric features and the cutting force in VAM. In this chapter, the geometric features of chip are introduced into the cutting force model to predict the cutting force in elliptical vibration machining (EVM) and VAMILL.

6.2 Elliptical Vibration Cutting

6.2.1 Elliptical Tool Path Dimensions

2-D VAM, also called elliptical vibration cutting (EVC), is the most common type of VAM used for turning. In EVC the tool oscillates in both the thrust and cutting directions [2]. An example of a 2-D tool path is shown in Figure 6.1.

6.2.2 Analysis and Modeling of EVC Process

6.2.2.1 Analysis and Modeling of Tool Motion

In the EVC process, the cutting tool is actuated by two actuators with a phase difference, and the tool motion is coupled to form an elliptical trajectory in space. The tool tip intermittently contacts the workpiece, as shown in Figure 6.1. It should be noted that the tool edge radius is assumed to be ideally sharp in this chapter.

Figure 6.1 shows two adjacent cutting cycles in the x–z plane. Time t_1 to t_2 indicates the previous cutting cycle, which constitutes the chip upper boundary, while time t_3 to t_4 indicates the present cutting cycle, which constitutes the chip bottom boundary. The tool trajectory expression established in prior research only considered the relationship of motion between the tool and the workpiece, and did not take into account the positional relationship between the vertical amplitude and nominal depth of cut. By fully taking into account the relationship between the tool and workpiece, the tool path can be expressed as

$$\begin{cases} x = a\cos(2\pi ft) - vt \\ z = c - [NDOC - c\sin(2\pi ft)] \end{cases} \tag{6.1}$$

where a is the transverse amplitude, c is the vertical amplitude, $NDOC$ is the nominal depth of cut, f is the vibration frequency, and v is the cutting speed.

6.2.2.2 Modeling of Chip Geometric Feature

It is very difficult to establish a model of the geometric features of the chip using a dynamic model, and so Eq. (6.1) is converted into a standard elliptical trajectory equation according to its motion characteristics. If the depth ratio (DR) is the ratio of $NDOC$ to c, then the

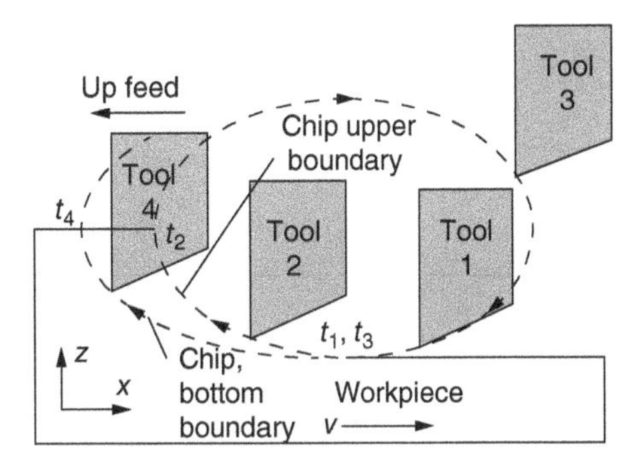

Figure 6.1 Elliptical vibration cutting.

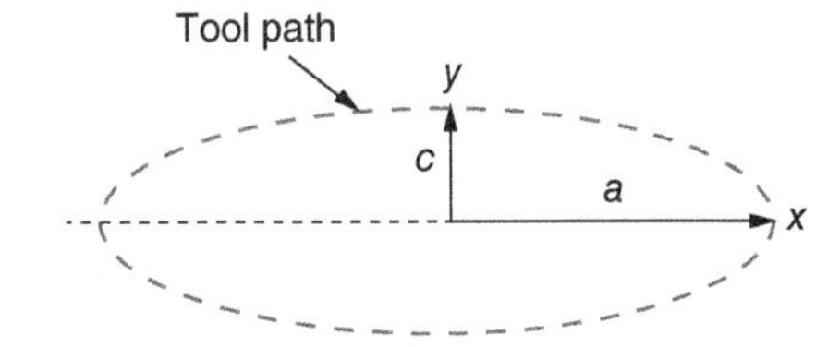

Figure 6.2 Tool path.

elliptical trajectory expression changes with *DR*, and in Figure 6.1 *DR* is equal to 1. Elliptical tool paths are described in terms of the major and minor axis dimensions as shown in Figure 6.2. The dimension a is half of the maximum length of the tool path in the cutting direction and c is half of the maximum dimension of the tool path in the thrust direction.

With a coordinate system that is centered on the tool path, dimensions a and c can be used along with the mathematical equation of an ellipse shown in Eq. (6.2) to approximate the tool path.

$$\frac{x^2}{a^2} + \frac{{z_1}^2}{c^2} = 1 \tag{6.2}$$

Since the two cycles are consecutive, the distance between the previous cycle and the present cycle is the movement of the tool along the cutting direction during a cycle time, namely an up feed per cycle. Therefore, the former cycle elliptical trajectory can be obtained as follows:

$$\frac{(x - ufpc)^2}{a^2} + \frac{{z_0}^2}{c^2} = 1 \tag{6.3}$$

where $ufpc = v/f$ and the subscripts 0 and 1 denote the previous and present cycles, respectively.

When $DR < 1$, if the workpiece is a brittle material such as ceramic, the chip produced is discontinuous, as shown in Figure 6.3a. However, if it is a plastic material, the chip is continuous. When $DR > 1$, if the workpiece is plastic, the chip produced is continuous, as shown in Figure 6.3b. When $DR > 1$ or $DR < 1$, the pervious elliptical trajectory moves along the z axis and the distance is *DF*, and so Eq. (6.4) can apply. The present elliptical trajectory moves along the x axis and the distance is *ufpc*, so that Eq. (6.5) can be obtained:

$$\frac{x^2}{a^2} + \frac{(z_1 \pm DF)^2}{c^2} = 1 \tag{6.4}$$

$$\frac{(x - ufpc)^2}{a^2} + \frac{(z_0 \pm DF)^2}{c^2} = 1 \tag{6.5}$$

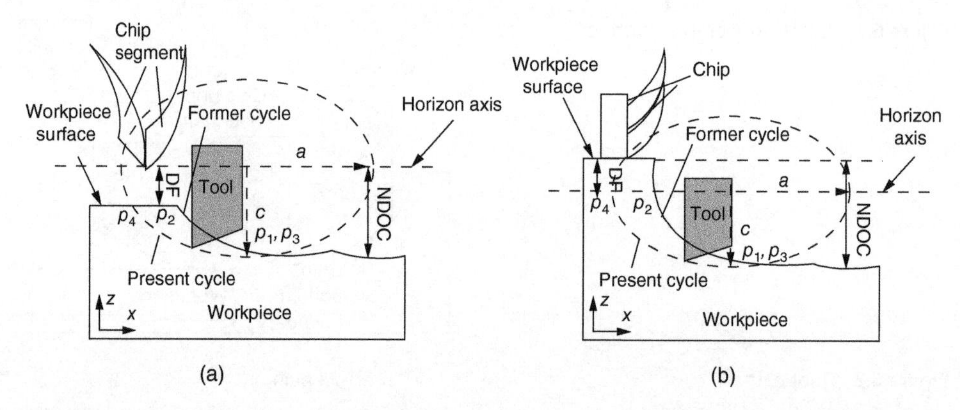

Figure 6.3 Positional relationship between the tool and the workpiece under different conditions. (a) $DR < 1$. (b) $DR > 1$.

where $DF = abs\,(NDOC - c)$, when $DR > 1$, DF is positive and $DR < 1$, when DF is negative.

If the former and present elliptical trajectory expressions are solved for the z-component under different DRs, the difference between the two expressions in the z-component is the transient chip thickness:

$$\text{chip}_{tr} = \begin{cases} z_1 - a \le x < -a + ufpc \\ z_0 - z_1 - a + ufpc \le x < ufpc/2 \qquad \text{(for } DR = 1\text{)} \\ 0 \qquad \text{otherwise} \end{cases} \tag{6.6}$$

$$\text{chip}_{tr} = \begin{cases} z_1 - a\sqrt{1 - DF^2/c^2} \le x < -a\sqrt{1 - DF^2/c^2} + ufpc \\ z_0 - z_1 - a\sqrt{1 - DF^2/c^2} + ufpc \le x < ufpc/2 \qquad \text{(for } DR < 1\text{)} \\ 0 \qquad \text{otherwise} \end{cases} \tag{6.7}$$

$$\text{chip}_{tr} = \begin{cases} -z_1 + DF - a \le x < -a + ufpc \\ z_0 - z_1 - a + ufpc \le x < ufpc/2 \qquad \text{(for } DR > 1\text{)} \\ 0 \qquad \text{otherwise} \end{cases} \tag{6.8}$$

The expression of transient chip thickness changes with DR. Figure 6.4 shows the variation in transient chip thickness under different values of DR. When the cutting speed is far greater than the vibration frequency, micro-dimples are formed on the workpiece. The sectional area of the chip in the x–z plane can be obtained by integrating the transient chip thickness in the definitional domain.

$$\text{Area}_{x-z} = \int_{-a}^{-a+ufpc} z_1 dx + \int_{-a+ufpc}^{ufpc/2} (z_0 - z_1) dx \qquad \text{(for } DR = 1\text{)} \tag{6.9}$$

$$\text{Area}_{x-z} = \int_{-a\sqrt{1-\frac{DF^2}{c^2}}}^{-a\sqrt{1-\frac{DF^2}{c^2}}+ufpc} z_1 dx + \int_{-a\sqrt{1-\frac{DF^2}{c^2}}+ufpc}^{ufpc/2} (z_0 - z_1) dx \qquad \text{(for } DR > 1\text{)} \tag{6.10}$$

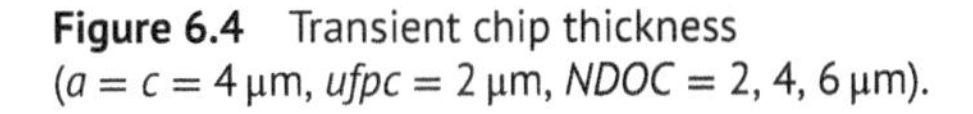

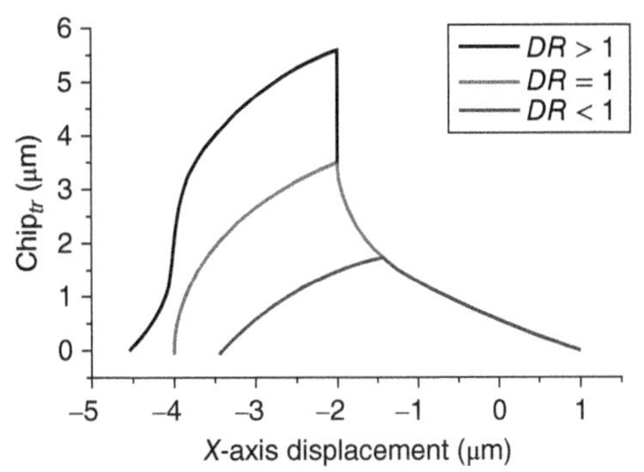

Figure 6.4 Transient chip thickness ($a = c = 4\,\mu\text{m}$, $ufpc = 2\,\mu\text{m}$, $NDOC = 2, 4, 6\,\mu\text{m}$).

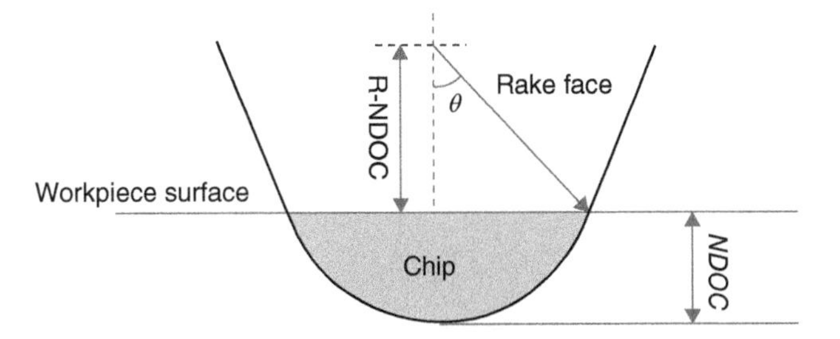

Figure 6.5 The tool area of y–z plane in micro-groove generation.

$$\text{Area}_{x-z} = \int_{-a}^{-a+ufpc} -z_1\,dx + \int_{-a+ufpc}^{ufpc/2} (z_0 - z_1)dx + \int_{-a+ufpc}^{-a\sqrt{1-\frac{DF^2}{c^2}}+ufpc} -z_1\,dx \qquad (\text{for } DR < 1) \tag{6.11}$$

Since the chip thickness in EVC varies, there must be a maximum value and this value occurs at p2 when the tool is in the present cycle according to Figure 6.2, and the calculation is given according to Eq. (6.12).

$$\text{chip}_{max} = \begin{cases} c\sqrt{1 - \dfrac{(a - ufpc)^2}{a^2}} & DR = 1 \\ c\sqrt{1 - \dfrac{\left(\dfrac{a\sqrt{c^2 - DF^2}}{c} - ufpc\right)^2}{a^2}} & DR > 1 \\ \sqrt{1 - \dfrac{(a - ufpc)^2}{a^2}} + DF & DR < 1 \end{cases} \tag{6.12}$$

The tool area of the y–z plane shows the contact conditions between the rake face and workpiece. In the EVC process, the cutting depth is much smaller than the tool radius, as shown in Figure 6.5.

According to the geometric relationship shown in Figure 6.5, θ can be expressed as follows:

$$\theta_{0,1} = \cos^{-1}\left(\frac{R - z_{0,1}}{R}\right) \tag{6.13}$$

where $\theta_{0,\,1}$ is half of the central angle for the previous and present cycles and R is the tool radius.

Then the tool area of the y–z plane at different depths of cut can be obtained:

$$A_{0,1} = R^2\theta_{0,1} - \frac{1}{2}(R - z_{0,1})R\sin\theta_{0,1} \tag{6.14}$$

Notice that the sectional area of the chip in the y–z plane is the difference between the tool areas of the former and present cycles at the same x-component:

$$\text{Area}_{y-z} = \begin{cases} A_1 & -a \le x \le -a + ufpc \\ A_0 - A_1 & -a + ufpc \le x \le ufpc/2 \\ 0 & \text{otherwise} \end{cases} \quad (\text{for } DR = 1 \text{ and } DR > 1) \tag{6.15}$$

$$\text{Area}_{y-z} \begin{cases} A_1 & -a\sqrt{1 - DF^2/c^2} \le x \le -a\sqrt{1 - DF^2/c^2} + ufpc \\ A_0 - A_1 & -a\sqrt{1 - DF^2/c^2} + ufpc \le x \le ufpc/2 \\ 0 & \text{otherwise} \end{cases} \tag{6.16}$$

According to the sectional areas of the chip in the x–z and y–z planes, together with the definitional domain, the volume of chip removed can be calculated under different DRs as follows:

$$V_{chip} = \begin{cases} \int_{-a}^{-a+ufpc} A_1 dx + \int_{-a+ufpc}^{ufpc/2} (A_1 - A_0) dx & DR = 1 \\ \int_{\frac{-a\sqrt{b^2 - DF^2}}{b}}^{\frac{-a\sqrt{b^2 - DF^2}}{b} + ufpc} A_1 dx + \int_{\frac{-a\sqrt{b^2 - DF^2}}{b} + ufpc}^{ufpc/2} (A_1 - A_0) dx & DR < 1 \\ \int_{-a}^{-a+ufpc} A_1 dx + \int_{-a+ufpc}^{ufpc/2} (A_1 - A_0) dx + \int_{-a+ufpc}^{-a\sqrt{1-DF^2/c^2} - ufpc} A_1 & DR > 1 \end{cases} \tag{6.17}$$

6.2.2.3 Modeling of Transient Cutting Force

According to the above equations, it can be concluded that the chip thickness in the EVC process is smaller than that in conventional cutting, which means that the shear angle in EVC is larger. Based on the cutting force model, which has already been verified in predicting the cutting forces [5], this section introduces the geometric features of the chip into the model in order to calculate the transient cutting force in EVC, which can be expressed as

$$F_c = \begin{cases} \text{Area}_{y-z}\tau_s \dfrac{\cos(\beta - \alpha)}{\sin\varphi_c \cos(\varphi_c + \beta - \alpha)} & \varphi = \varphi_c \\ \text{Area}_{y-z}\tau_s \dfrac{\cos(\beta - \alpha)}{\sin\varphi_{evc} \cos(\alpha + \beta - \varphi_{evc})} & \end{cases} \tag{6.18}$$

where τ_s, the flow stress, is relevant to shear angle, β is the mean friction angle, α is the rake angle, and φ_{evc} and φ_c are the shear angles in the conventional cutting zone and friction reversal zone, respectively.

According to the principle of minimum energy, the shear angle will assume such a value as to minimize the total work done. Therefore, it can be obtained as follows:

$$\begin{cases} \dfrac{dF_c}{d\varphi_c} = -\text{Area}_{y-z}\tau_s \cos(\beta - \alpha) \dfrac{\cos\varphi_c \cos(\varphi_c + \beta - \alpha) - \sin\varphi_c \sin(\varphi_c + \beta - \alpha)}{\sin^2\varphi_c \cos^2(\varphi_c + \beta + \alpha)} = 0 & \varphi = \varphi_c \\ \dfrac{dF_c}{d\varphi_{evc}} = \text{Area}_{y-z}\tau_s \cos(\beta + \alpha) \dfrac{\cos\varphi_{evc} \cos(\beta + \alpha - \varphi_{evc}) + \sin\varphi_{evc} \sin(\beta + \alpha - \varphi_{evc})}{\sin^2\varphi_{evc} \cos^2(\beta + \alpha - \varphi_{evc})} = 0 & \varphi = \varphi_{evc} \end{cases} \tag{6.19}$$

Thus,

$$\begin{cases} \varphi_c = \dfrac{\pi}{4} - \dfrac{\beta}{2} + \dfrac{\alpha}{2} \\ \varphi_{evc} = \dfrac{\pi}{4} + \dfrac{\beta}{2} + \dfrac{\alpha}{2} \end{cases} \tag{6.20}$$

According to the conventional thin shear plane model, flow stress can be expressed as

$$\tau_s = \begin{cases} \dfrac{R_f \cos(\varphi_c + \tan^{-1}(\cos(\pi/2 - 2\varphi_c)/\sin(\pi/2 - 2\varphi_c)))}{2\sqrt{R^2 - (R - \text{chip}_{tr})^2}(\text{chip}_{tr}/\sin(\varphi_c))} & \varphi = \varphi_c \\ \dfrac{R_f \cos(\varphi_{evc} + \tan^{-1}(\cos(2\varphi_{evc} - \pi/2)/\sin(2\varphi_c - \pi/2)))}{2\sqrt{R^2 - (R - \text{chip}_{tr})^2}(\text{chip}_{tr}/\sin(\varphi_{evc}))} & \varphi = \varphi_{evc} \end{cases} \tag{6.21}$$

where R_f is the resultant force.

By combining Eqs. (6.18, 6.21), we can conclude that the components of cutting force consist of the sectional area of the chip, the flow stress, and the shear angle. Before carrying out an EVC experiment, a cutting tool has been chosen, which means that the tool radius and rake angle are fixed and the workpiece has also already been chosen so that its properties are fixed too. In the EVC process, the flow shear stress is relative to the shear angle and once the cutting parameters are determined, the shear angle is fixed. Only the sectional area of the chip in the y–z plane is variable and can be obtained; furthermore, this depends on the transient chip thickness. In Eq. (6.18), the sectional area of the chip in the y–z plane depends on chip thickness as an independent variable and the rest as a cutting force coefficient, which can be obtained by collecting cutting force data. Once the cutting force coefficient is obtained, the transient cutting force can be calculated from Eq. (6.18). In general, it is difficult to use conventional methods to calculate the cutting force, and it takes a long time to do so. However, the method proposed in this section is easy to apply and the transient cutting force is obtained quickly. What is important is that by establishing

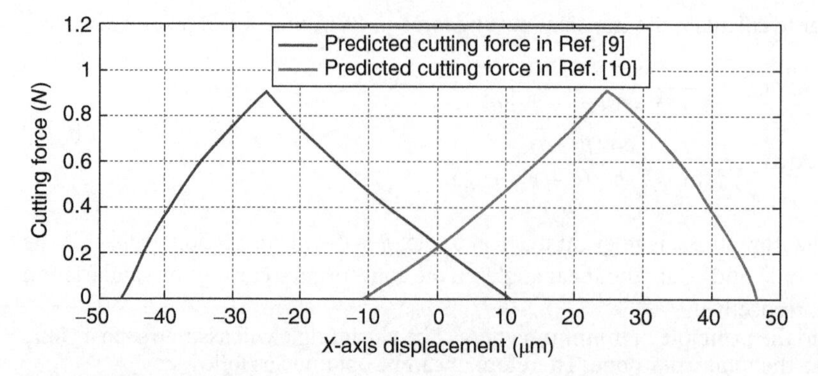

Figure 6.6 Comparison of predicted cutting force in one cycle for $DR < 1$ ($a = 47.5$ μm, $c = 7.3$ μm, $ufpc = 21.638$ μm, $NDOC = 6$ μm).

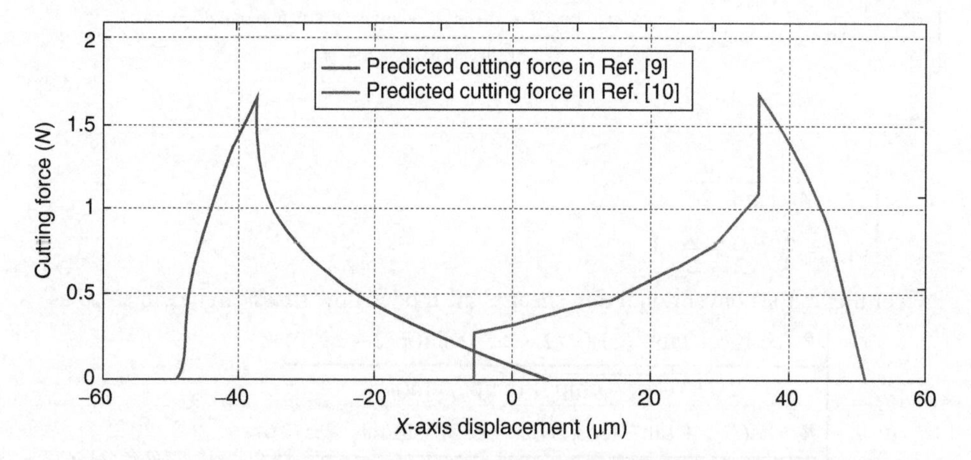

Figure 6.7 Comparison of predicted cutting force in one cycle for $DR > 1$ ($a = 47.5$ μm, $c = 7.3$ μm, $ufpc = 21.638$ μm, $NDOC = 9$ μm).

the relationship between the geometric features of the chip and the cutting force, we can infer the effects of cutting parameters such as amplitude and vibration frequency on the performance of cutting force.

6.2.3 Validation of the Proposed Method

Lin et al. [9] validated their proposed method for predicting transient cutting force, and the improved cutting force model was compared to a previous cutting force model [10]. The conventional cutting force model in the literature consists of shear angles, a hardness multiplier, and the sectional area of the chip. Before calculating the cutting force, all of these parameters should first be calculated, and then substituted in the cutting force model, before finally plotting the cutting force lines. The calculated cutting force coefficient by collecting values of predicted cutting force from the literature and plotting the cutting force lines in one cycle are shown in Figures 6.6 and 6.7.

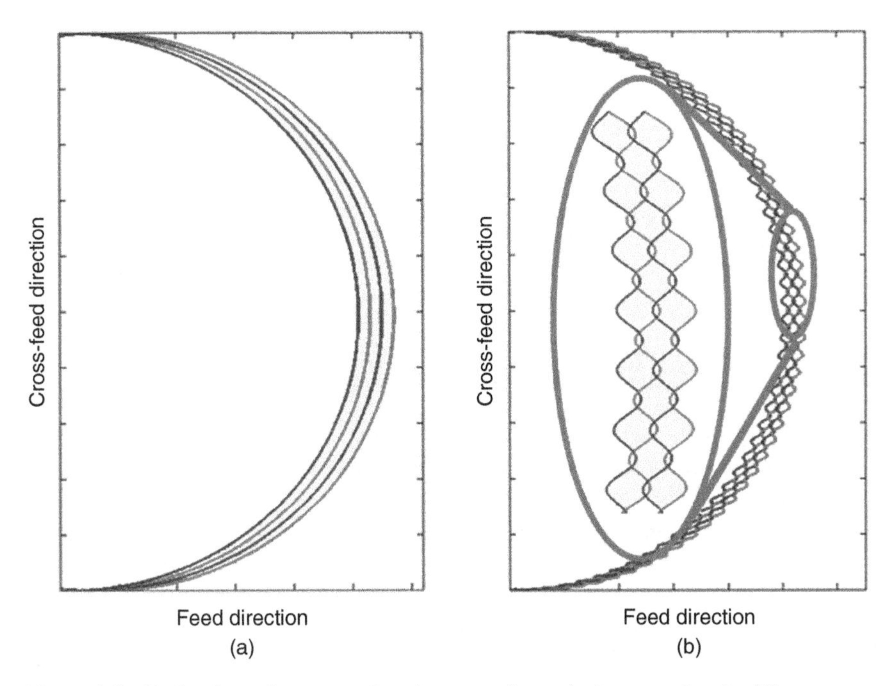

Figure 6.8 Tool trajectories comparison between the typical conventional milling process and vibration-assisted milling process.

It should be noted that the cutting direction in this section is the opposite of the previous study so that the two predicted cutting force lines are symmetrical. The predicted cutting force was collected and substituted into Eq. (6.18) to calculate the cutting force coefficient [10]. The results show that when $DR < 1$, the cutting force coefficient is 8.06e-5, and when $DR > 1$, the cutting force coefficient is 1.0194e-3. Once the cutting force coefficient is obtained, the cutting force at any position of the workpiece can be calculated from Eq. (6.18). When $DR < 1$, the error between the predicted cutting force in this study and the previously predicted cutting force in Ref. [10] is 1.16%. When $DR > 1$, the error is 1.94%. As shown in Figure 6.8, the predicted cutting force found here is somewhat different from that found before. At the beginning of cutting, the predicted cutting force in the literature first sharply increases and then increases steadily; however, according to Figure 6.4 the transient chip thickness increases smoothly, so that the cutting force should also increase smoothly, and the predicted cutting force in this paper should be more reasonable. As the cutting process is underway, there is a sudden increment in cutting force due to a sudden increment in chip thickness and when the material of the workpiece is brittle increased cutting force will damage the workpiece surface.

6.3 Vibration-Assisted Milling

Having discussed the cutting force model for EVC, it should be noted that the cutting force model for VAMILL is more complex. So far, the tool–workpiece separation (TWS) mechanism, which has a significant effect on the cutting process, has not been systematically investigated for VAMILL. An accurate cutting force model considering various TWS conditions is urgently needed to allow the investigation of the underlying cutting mechanism in the VAMILL process. This section systematically investigates the TWS mechanism and establishes a cutting force prediction model for VAMILL based on the instantaneous uncut chip thickness considering the TWS. Meanwhile, the effects of machining and vibration parameters on cutting forces are numerically investigated. Finally, a generic cutting force model for VAMILL is established, offering guidance for not only the determination of optimal machining and vibration parameter sets but also further investigation of the cutting mechanism.

6.3.1 Tool–Workpiece Separation in Vibration Assisted Milling

A two-dimensional VAMILL is investigated. The vibrations are applied simultaneously in both the feed (x) and the cross-feed (y) directions. The results and conclusion of the present study are also applicable to one-dimensional VAMILL, i.e. the vibration is applied in either the feed or cross-feed direction. The relative displacement (x_i, y_i) of the tool tip to workpiece in VAMILL can be described as

$$\begin{cases} x_i = ft + r\sin\left[\omega t - \dfrac{2\pi(z_i - 1)}{Z}\right] + A\sin(f_x + \emptyset_x) \\ y_i = r\cos\left[\omega t - \dfrac{2\pi(z_i - 1)}{Z}\right] + B\sin(f_y + \emptyset_y) \end{cases} \tag{6.22}$$

where r and ω are the radius and angular velocity of the cutter, z_i is the ith cutter tooth respectively, and Z is the number of flutes, f is the feed velocity, A and B are the vibration amplitudes, f_x and f_y are the vibration frequencies, and ϕ_x and ϕ_y are the phase angles, in the x- and y-directions, respectively.

From Eq. (6.22), it can be noted that in two-dimensional VAMILL, except for t, there are 10 independent parameters required to determine the tool tip trajectory. Owing to the effects of so many parameters, the tool tip trajectory in VAMILL is much more complex. To investigate the material removal mechanism, it is necessary to explicitly obtain the uncut chip thickness, despite its difficulties.

Figure 6.8 compares the trajectories in the conventional and vibration-assisted milling processes. It can be noted that in conventional milling, the instantaneous uncut chip thickness is usually determined based on two consecutive tool paths, and the influence of the workpiece material removed by previous tool paths is generally neglected. When considering the vibrations applied on the machining system, the milling process becomes highly intermittent. Cutting edges are not always in contact with the workpiece during one tool passing period. The chip generated by one cutting tooth is also intermittent even in one tool edge passing period; therefore, the uncut chip thickness is determined according to the maximum contour formed by all previous cutting trajectories and current cutting trajectory. A closed-form analytical solution of instantaneous uncut chip thickness may not exist for VAMILL, so numerical simulations are carried out to describe this complex and intermittent process of chip formation.

In VAM machining, the TWS potentially occurs during the cutting process, as shown in Figure 6.9a. In the TWS regions, the instantaneous uncut thickness changes to zero, and the

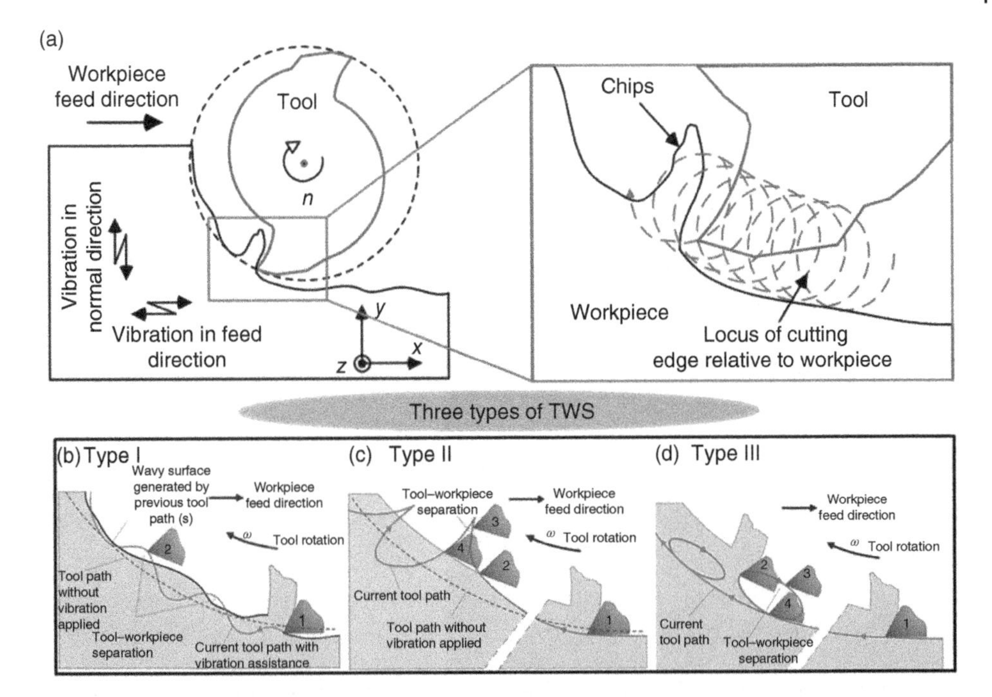

Figure 6.9 Schematics of three types of tool–workpiece separation.

TWS is recognized as the main reason for the benefits of VAM. Therefore, the investigation of the mechanism and the conditions of TWS are of significance for the modeling of cutting force and understanding of the cutting mechanism in VAM. Based on the analysis of the kinematics of the VAMILL, three types of TWS are proposed, as shown in Figure 6.9b–d.

TWS type I: as shown in Figure 6.9b, TWS type I separation occurs when the current tool path obtained with the aid of vibration partly overlaps with the surface contour left by previous cutting path(s). In these regions of overlapping, the cutting tool edge may lose contact with the workpiece and discontinuous chips are generated. As part of the material in the current cutting path has been removed by previous cutting path(s), periodical TWS takes place.

Generally, TWS type I could occur only when the following parametric conditions are satisfied: the vibration is applied in the feed direction, and vibration frequency is not the integral times of the spindle rotation frequency and vibration amplitude is at least greater than half of the feed per tooth.

When the vibration frequency is an even multiple of the spindle rotation frequency, the phase difference between the adjacent cutting paths is zero. Here, the wave peaks (troughs) of the kth tool tip trajectory overlap with the peaks (troughs) of the $(k+1)$th tool tip trajectory, as shown in Figure 6.10a. Thus, no separation occurs.

When the vibration frequency is an odd multiple of the spindle rotation frequency, the phase difference between the adjacent cutting paths is π, which means that the wave peaks (troughs) of the kth tool tip trajectory overlap with those of the $(k+1)$th tool tip trajectory,

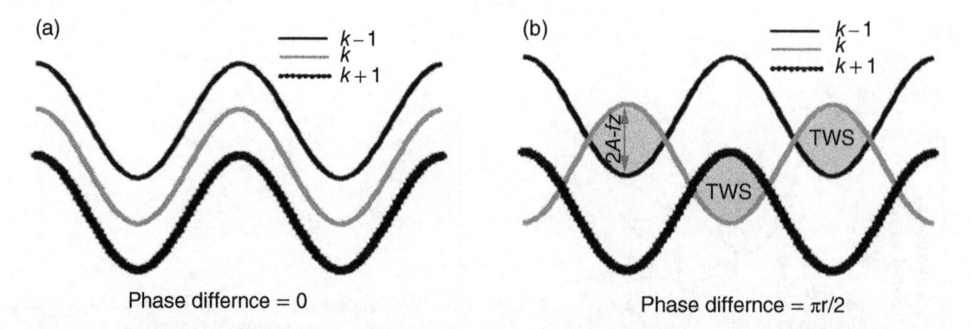

Figure 6.10 Tool tip trajectory of VAMILL. (a) Vibration frequency is an even multiple of spindle rotation frequency; (b) vibration frequency is an odd multiple of the spindle rotation frequency.

as shown in Figure 6.10b. In such a case, if the vibration amplitude is more than half of the feed per tooth, periodic separation will occur.

The sufficient condition for TWS type I is given as follows:

$$\begin{cases} 60f_x/n = \text{odd} \\ A > f_z/2 \end{cases} \tag{6.23}$$

TWS type II is shown in Figure 6.9c; TWS type II separation occurs in the current tool path. It can be seen that when vibration displacement in the tool's radial direction is larger than the instantaneous uncut chip thickness, the tool is cut out from the workpiece, thereby leading to the TWS. In this condition, the instantaneous uncut chip thickness can be expressed by

$$h_{dv} = f_z \sin\theta - x_w \sin\theta - y_w \cos\theta \tag{6.24}$$

where h_{DV} is the instantaneous uncut thickness, and x_w and y_w are the instantaneous displacements of the workpiece, respectively.

The sufficient condition for type II separation is $h_{DV} < 0$.

TWS type III: as shown in Figure 6.9d, when the direction of the component of the relative velocity between tool and workpiece in the cutting direction (i.e. tangential component) is opposed to the direction of tool rotation, the tool tip is behind the workpiece, thereby leading to the occurrence of TWS type III.

A velocity ratio K is defined in VAMILL as

$$K = \frac{\omega r}{v_w} \tag{6.25}$$

where v_w denotes the component of the velocity of workpiece in the instantaneous cutting direction, which can be computed as

$$v_w = 2\pi A f_x \cos(2\pi f_x t + \phi_x)\cos\theta + 2\pi B f_y \cos(2\pi f_y t + \phi_y)\sin\theta \tag{6.26}$$

When $K < 1$, the cutter is behind the workpiece, and TWS type III takes place. Conversely, when $K \geq 1$, the material will be removed by the cutter, and no TWS occurs.

Therefore, the preconditions for the occurrence of TWS type III is $K < 1$.

Figure 6.11 FE model for VAMILL.

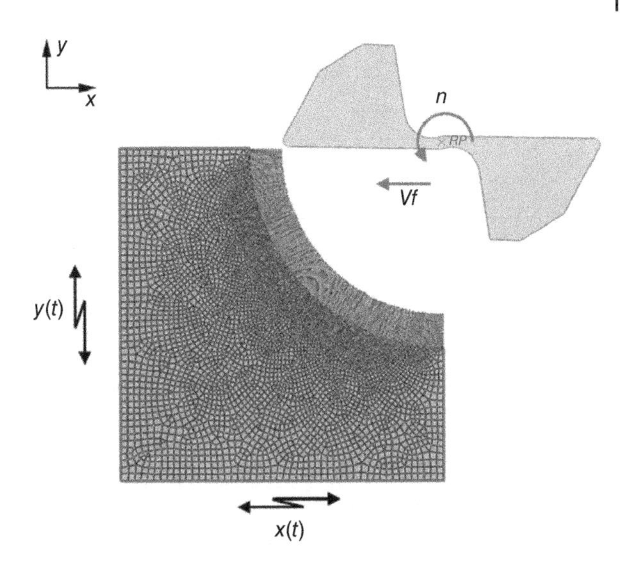

6.3.2 Verification of Tool–Workpiece Separation

Considering the high frequency and small amplitude of the vibration applied in VAMILL, the TWS time is usually of the order of 10^{-5} seconds. Thus, it is difficult to observe TWS experimentally. Finite element (FE) simulation has proved to be an effective tool in investigating the cutting process. The effects of geometric parameters (kinematics) and physical/mechanical properties of the cutting tool and workpiece can be considered in a FE cutting simulation. Moreover, the FE simulations could also be applied to studying cutting phenomena such as chip formation, minimum chip thickness, and temperature in the cutting zone, which is hard to observe experimentally.

To clearly investigate the time evolution of the workpiece–tool contact status in the VAMILL process, an FE model for VAMILL was established using the commercial package, ABAQUS/Explicit, as shown in Figure 6.11. On this basis, the proposed kinematic model and three types of TWS are verified.

AISI 1045 steel was chosen as the workpiece material, due to its popularity in the plastic injection molding industry. In order to reduce the computational time required, the cutter was set as a rigid body. The edge radius of the tool is 3 μm, and minor edge angle is 5°. The Johnson–Cook (JC) material model and Johnson–Cook damage model based on the nonlinear temperature and strain rate are used to describe the workpiece material behavior.

The primary equation for the JC material model describes the flow stress as

$$\sigma_y = [A + B(\varepsilon_p)^n][1 + Cln(\dot{\varepsilon}_p^*)][1 - (T^*)^m] \tag{6.27}$$

$$\dot{\varepsilon}_p^* = \frac{\dot{\varepsilon}_p}{\dot{\varepsilon}_{p0}} \tag{6.28}$$

$$T^* = \frac{T - T_0}{T_m - T_0} \tag{6.29}$$

where ε_p is the effective plastic strain, and $\dot{\varepsilon}_p$and $\dot{\varepsilon}_{p0}$are the plastic strain rate and effective plastic strain rate used for the calibration of the model, respectively. T and T_0 are the current

Table 6.1 Mechanical properties and parameters for AISI 1045 steel.

Properties and parameters	Notation	Value
Density	ρ	7800 kgm^{-3}
Thermal conductivity	K	38 Wm^{-1} K^{-1}
Specific heat	C	420 JKg^{-1} K^{-1}
Tayor–Qulnney coefficient	β	0.9
Initial yield stress	A	553 MPa
Hardening modulus	B	600 MPa
Strain rate dependency coefficient	C	0.0134
Work-hardening exponent	n	0.234
Thermal softening coefficient	m	1
The reference stain rate	$\dot{\varepsilon}_{p0}$	1 s^{-1}
Room temperature	T_0	300 K
Melting temperature	T_m	1733 K

Table 6.2 Machining and vibration-assisted parameters.

Set No.	Spindle speed (rpm)	Feed per tooth (μm)	*x*- direction vibration amplitude (μm)	*x*-direction vibration frequency (Hz)	*y*–direction vibration amplitude (μm)	*y*-direction vibration frequency (μm)	Phase difference
1	6000	20	5	1500	5	1500	π/2
2	6000	10	7	1500	7	1500	π/2
3	6000	20	5	9600	5	9600	π/2

and reference temperatures, respectively. The detailed parameters adopted in the FE model are given in Table 6.1.

To verify the three types of TWS proposed in Section 6.3.1, three sets of FE simulations were carried out and the corresponding machining and vibration parameters are listed in Table.6.2. Herein, the simulation Set 1 is carried out to illustrate the case that no TWS occurs; the simulation Set 2 is conducted to verify the TWS type I and II, and simulation Set 3 is applied to illustrate TWS type III.

From Figure 6.12, it can be seen that for Set 1 milling process simulation, the instantaneous uncut thickness is continuous in one cutting path and the chips are approximately consistent with those formed in the conventional milling. No TWS occurs in the machining process.

Figure 6.13a) shows the machined surface after multiple cutting paths. It can be observed that due to the overlap of the current and previous cutting path(s), periodical TWS occurs, thereby verifying the proposed TWS type I in VAMILL. As the periodic separation occurs

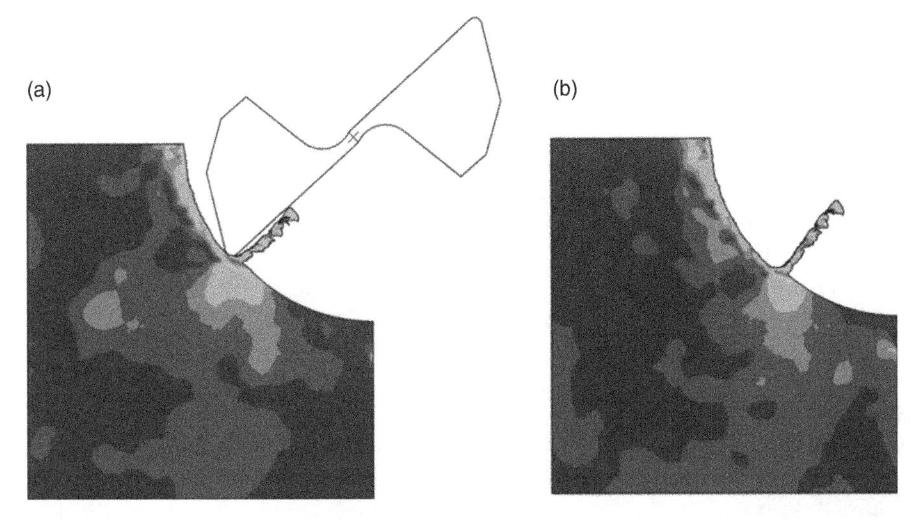

Figure 6.12 First set simulation results.

during the machining process, the cutting path becomes highly discontinuous; thus, the "block type" chips are formed, as shown in Figure 6.13b.

In Figure 6.13c, it can be noted that in the current cutting path, owing to the effects of the vibration applied on the workpiece, the cut-in and cut-out of tool continually occur in the direction of instantaneous uncut thickness, leading to the separation of tool–workpiece, which corresponds to TWS types I and II. Meanwhile, due to the lower separation frequency in TWS type II than in TWS type I, "C type" chips are formed, as shown in Figure 6.13d. It can be seen that "C type" chips are longer than the "block type" chips.

Figure 6.14a clearly shows the process of the occurrence of TWS type III. It can be observed that the cutting tool edge regains and loses contact with the workpiece material, due to the workpiece vibrations in the cutting direction. Although the tool and the workpiece will be separated, the chips are still not cut off by the tool, and finally continuous chips are formed, as shown in Figure 6.14b.

6.3.3 Cutting Force Modeling of VAMILL

Given the complexity of the tool trajectories in the VAMILL, it is difficult to derive a closed-form analytical solution for instantaneous uncut chip thickness. In addition, the TWS further exacerbates the complexity of the modeling of transient cutting thickness. In this study, a numerical simulation model for instantaneous uncut chip thickness in VAMILL is proposed, and on the basis of the proposed model a generic cutting force model of VAMILL is established.

6.3.3.1 Instantaneous Uncut Thickness Model

Figure 6.15 illustrates a typical cutter trajectory in VAMILL. Assuming that $O(x,y)$ are the coordinates of the tool center at any moment of time t, $A(x, y)$ are the coordinates of the tool tip, $\overrightarrow{OA}$ represents the vector from $O(x,y)$ to $A(x,y)$, and the intersections of $\overrightarrow{OA}$ (and its extension line) with the previous tool trajectories are recorded as $A_1, A_2, \ldots A_n$, respectively.

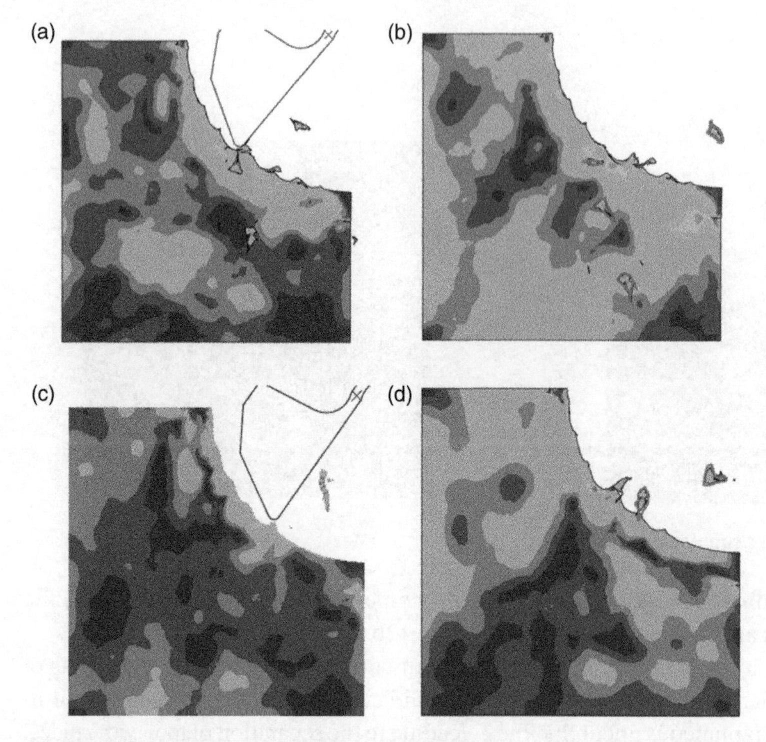

Figure 6.13 Second set simulation results.

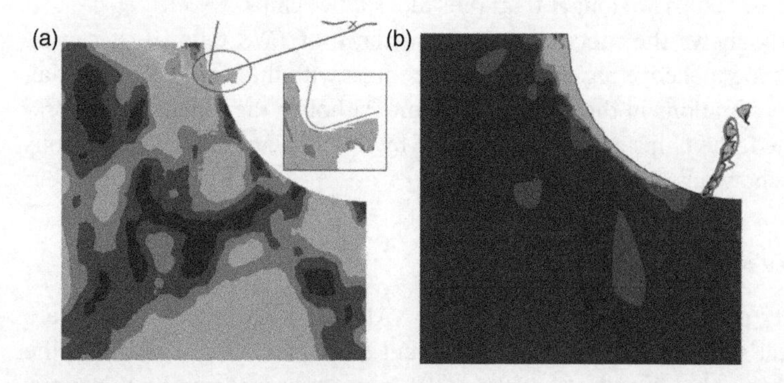

Figure 6.14 Third set simulation results.

If the direction of any one of $\overrightarrow{A_1A}, \overrightarrow{A_2A}, \ldots \overrightarrow{A_nA}$ is opposite to that of $\overrightarrow{OA}$, the tool tip is within the region where the workpiece material has been removed by the previous cutting path, and the instantaneous uncut thickness drops to zero.

Conversely, if all of $\overrightarrow{A_1A}, \overrightarrow{A_2A}, \ldots \overrightarrow{A_nA}$ have the same direction as $\overrightarrow{OA}$,the tool tip is in direct contact with the workpiece, i.e. no TWS, and the instantaneous uncut thickness is $\min\{|\overrightarrow{A_1A}|, |\overrightarrow{A_2A}|, \ldots |\overrightarrow{A_nA}|\}$.

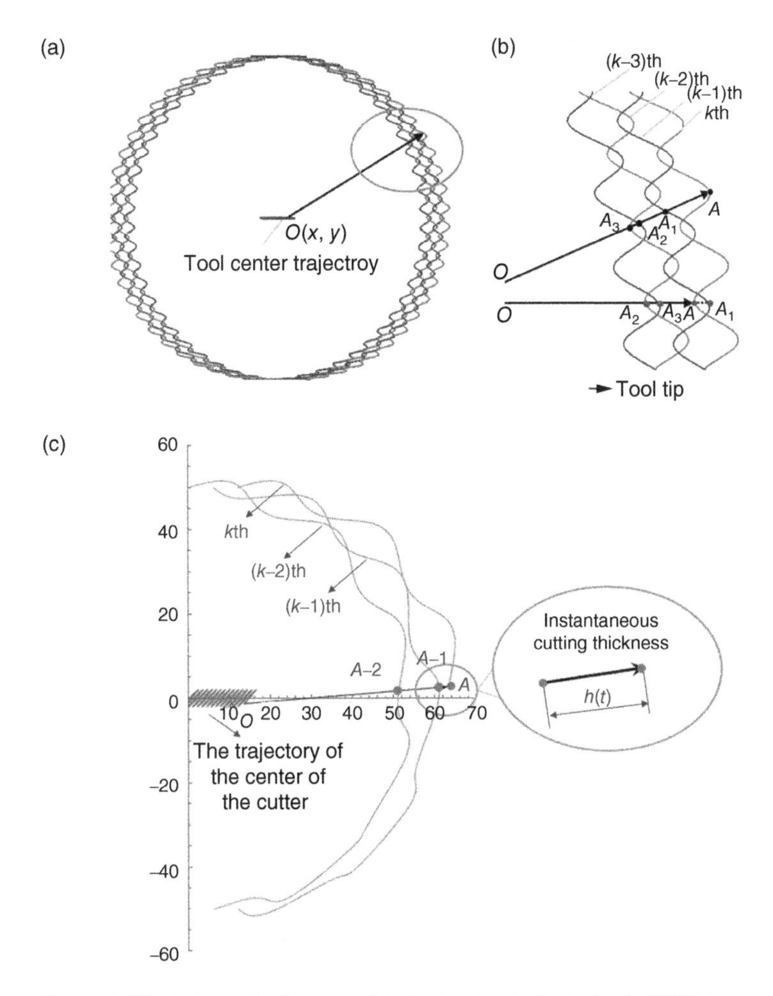

Figure 6.15 Schematic diagram of typical cutter trajectories in VAMILL.

Thus, the instantaneous uncut thickness can be expressed as (Figure 6.16)

$$h(t) = \begin{cases} \min\{|\overrightarrow{\mathrm{A_1A}}|, |\overrightarrow{\mathrm{A_2A}}| \cdots |\overrightarrow{\mathrm{A_nA}}|\}, & \text{same direction with } \overrightarrow{\mathrm{OA}} \\ 0, & \text{else} \end{cases} \tag{6.30}$$

Equation (6.30) can be used to determine the occurrence of TWS types I and II. However, as shown in Figure 6.8, $\overrightarrow{\mathrm{A_1A}}, \overrightarrow{\mathrm{A_2A}}, \ldots \overrightarrow{\mathrm{A_nA}}$ are found to have the same direction, and according to Eq. (6.30), it can be determined that no TWS occurs during the machining process. However, the position of the cutter at the time $t_{(i+2)}$ is behind those at the time of $t_{(i\mp1)}$ and t_i, indicating that the tool is behind the workpiece, and TWS type III occurs. Thus Eq. (6.30) cannot be applied to determine the occurrence of TWS type III. To solve this problem, the separation conditions in TWS type III can be described by

$$2\pi Af_x \cos(2\pi f_x t + \phi_x)\cos\theta + 2\pi Bf_y \cos(2\pi f_y t + \phi_y)\sin\theta \geq \omega r \tag{6.31}$$

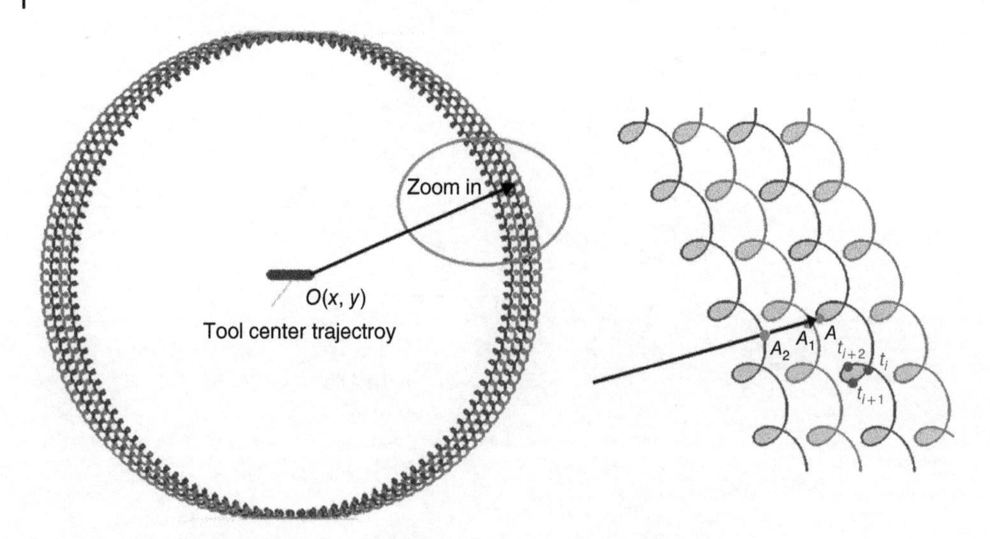

Figure 6.16 Schematic diagram of the other typical cutter trajectories.

Thus, the instantaneous uncut thickness can be further depicted as

$$h(t) = \begin{cases} \min\{|\overrightarrow{A_1A}|, |\overrightarrow{A_2A}|, \ldots |\overrightarrow{A_nA}|\}, \text{determining condition} \\ 0, \quad \text{else} \end{cases} \tag{6.32}$$

Herein, the determining condition refers to the case where $\overrightarrow{A_1A}, \overrightarrow{A_2A}, \ldots \overrightarrow{A_nA}$ have the same directions, and $2\pi Af_x \cos(2\pi f_x t + \phi_x)\cos\theta + 2\pi Bf_y \cos(2\pi f_y t + \phi_y)\sin\theta < \omega r$.

6.3.3.2 Cutting Force Modeling of VAMILL

Given the complexity of the instantaneous uncut thickness in VAMILL, the cutting force model proposed in the previous study [6], which was based on an analytical model used to calculate the instantaneous uncut chip thickness, is no longer applicable in VAMILL modeling (Figure 6.17). Ding et al. [1] proposed a numerical model to study the kinematics in VAMILL by considering the TWS.

In this study, based on the proposed instantaneous uncut thickness model, an improved cutting force model for VAMILL is established, as shown in Figure 6.9. In the milling process, the tangential force (F_t) and the radial force (F_r) acting on a differential flute can be depicted as

$$\begin{bmatrix} F_t \\ F_r \end{bmatrix} = a_p \left(\begin{bmatrix} K_t \\ K_r \end{bmatrix} h(t) + \begin{bmatrix} K_{te} \\ K_{re} \end{bmatrix} \right) g(\emptyset_j(t)) \tag{6.33}$$

where K_t and K_r are the tangential and radial cutting coefficients, respectively. a_p is the axial cutting depth. K_{te} and K_{re} are the tangential and radial of the cutting edge coefficients, respectively. $g(\emptyset_j(t))$ is the window function that determines whether a tooth is in or out. The tooth is in if $\emptyset_{st} \leq \emptyset_j\,(t) \leq \emptyset_{ex}$, where $\emptyset_{st}$ and $\emptyset_{ex}$ are the starting and exiting angles, respectively. This function is given by

$$g(\emptyset_j(t)) = \begin{cases} 1, \emptyset_{st} \leq \emptyset_j(t) \leq \emptyset_{ex} \\ 0, \text{else} \end{cases} \tag{6.34}$$

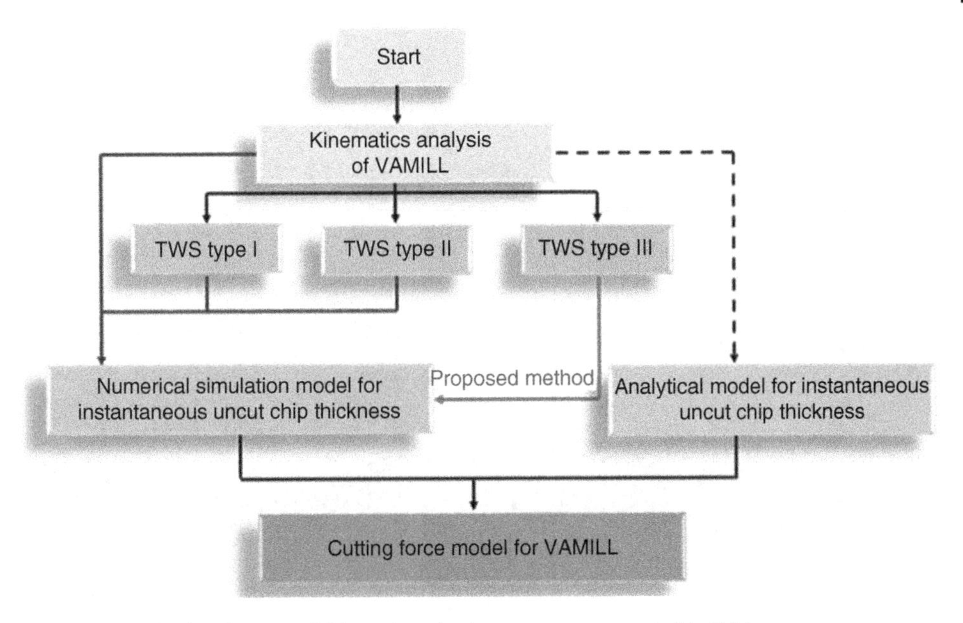

Figure 6.17 Cutting force model based on the instantaneous uncut chip thickness.

Therefore, the cutting forces on a cutter tooth in the x and y directions can be expressed as

$$\begin{bmatrix} F_x \\ F_y \end{bmatrix} = \begin{bmatrix} -\cos(\emptyset_j(t)) & -\sin(\emptyset_j(t)) \\ \sin(\emptyset_j(t)) & -\cos(\emptyset_j(t)) \end{bmatrix} \begin{bmatrix} F_t \\ F_r \end{bmatrix} \tag{6.35}$$

6.3.4 Discussion of Simulation Results and Experiments

Three typical sets of machining experiments were carried out on a three-axis precision milling machine tool (NANOWAVE MTS5R), and a novel 2D vibration stage with high stiffness (50 N/μm) and wide vibration frequency (9 kHz) was designed. Figure 6.18 illustrates a schematic of the vibration stage, which is driven by two piezo actuators (P-844.20) that are vertical to each other; the control signals are set by a host computer and amplified through a high voltage piezo amplifier, and then are used to drive the piezo actuators. Meanwhile, the stage displacement data is detected by two high precision capacitive sensors (CS005, Micro-epsilon) and fed back to the host computer through data acquisition cards for recording. The motion error of 0.25 μm is measured for the vibration stage used in this experiment.

The machine tool is equipped with three precision linear stages that are driven by DC servo motors with the smallest feed of 0.1 μm, and a high-speed spindle with a speed range from 5000 to 80 000 rpm. A typical experimental setup is presented in Figure 6.18b. A three-component piezoelectric dynamometer (Kistler 9256C2) was mounted on the Y stage to measure the feed and cross-feed cutting forces. The workpiece is clamped on a fixture attached to the X–Z stages (Figure 6.18(a)). Two-flute carbide micro-flat-end mills with a diameter of 1 mm were used in slot milling with AISI 1045. In order to verify the proposed

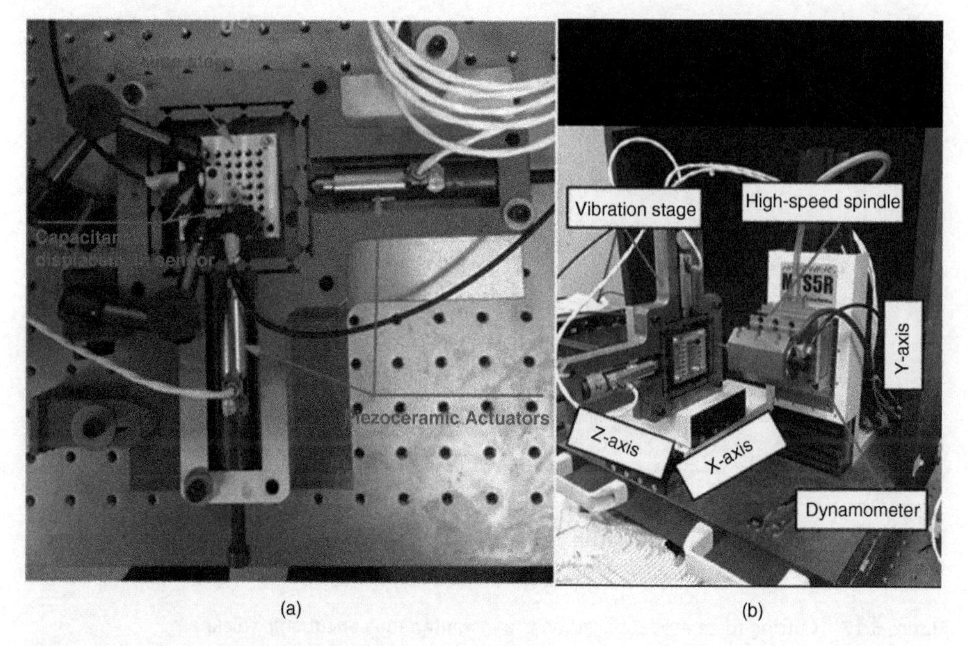

Figure 6.18 Schematic diagram and experimental setup of the VAMILL system. Source: Courtesy of Dehong Huo, Wanqun Chen, and Lu Zheng.

cutting force model, the machining parameters used in the machining experiments were the same as those used in the simulation (see Table 6.2).

Figure 6.19 compares the tool trajectory (a, c, and e) and the instantaneous uncut thickness (b, d, and f) in each set of machining experiments. In the first set of experiments, the vibration applied in the machining system induced a fluctuation in the instantaneous uncut thickness (see Figure 6.19b), compared with the conventional milling process. But no TWS occurs during the machining process.

In the second set (as shown in Figure 6.19d), owing to the effects of the vibration applied in the machining system, the troughs (peaks) of the current cutter trajectory intersect with the peaks (troughs) of the following cutter trajectory, further causing the periodic separation of tool and workpiece (TWS type II). In the third set, neither TWS type I nor II is found in the tool trajectories. However, according to Eq. (6.35), the component of the vibration velocity of the workpiece in cutting direction is larger than the instantaneous velocity of the cutter; as a result, the tool tip is behind the workpiece and TWS type III takes places. Meanwhile, in some region, the instantaneous uncut thickness drops to zero, as shown in Figure 6.19f.

Figures 6.20–6.22 compare the cutting force profiles obtained in each set of simulation and experiments. It can be seen in Figure 6.20 that in the first set of experiments and simulations, owing to the application of assisted vibration, the cutting force exhibits evident fluctuations in one tool passing cycle, which is different from that obtained from conventional milling. In the second set of experiments and simulations, as shown in Figure 6.13, the cutting force is conversely periodically intermittent, which can be attributed to the effects of the proposed TWS. Besides, it can be seen that the cutting forces obtained in the third set of simulations and experiments are intermittent in the cut-out region, which can be applied to validate the occurrence of TWS type III.

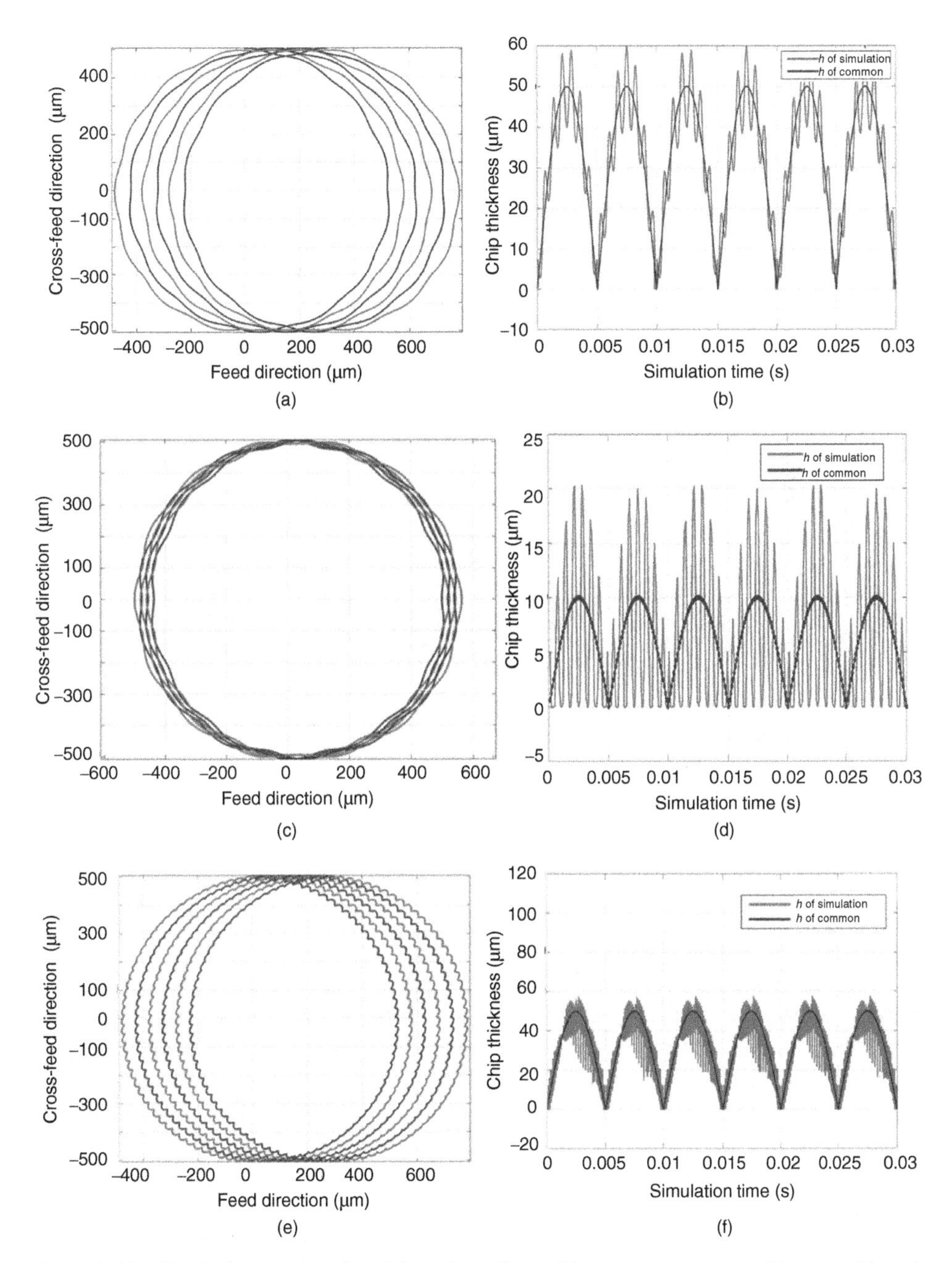

Figure 6.19 Simulation results of tool tip trajectories and instantaneous uncut thickness. (a) tool trajectory in set 1; (b) instantaneous uncut thickness in set 1; (c) tool trajectory in set 2; (d) instantaneous uncut thickness in set 2; (e) tool trajectory in set 3; (f) instantaneous uncut thickness in set 3.

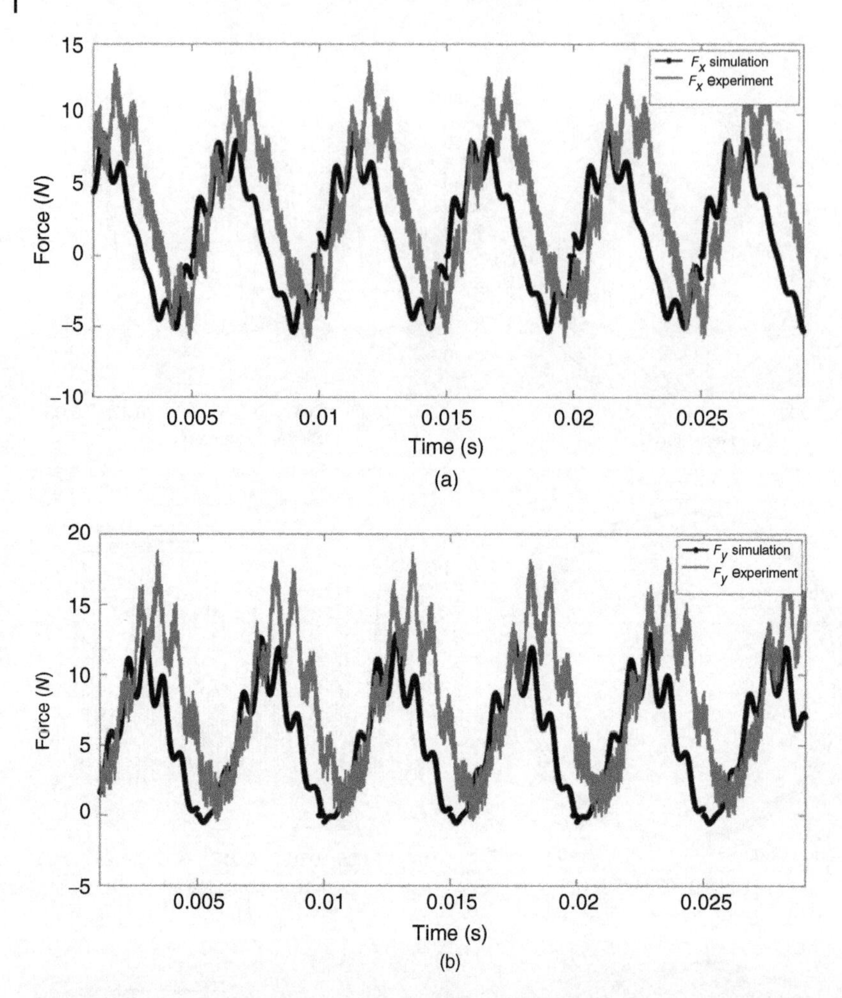

Figure 6.20 Cutting force of 1st set simulation and experiment. (a) x direction; (b) y direction.

The comparisons between the simulated and experimental cutting force in the global coordinate systems (x and y) in the three tests (set 1, 2, and 3) are given in Figures 6.20–6.22, respectively. As shown in the Figures 6.20–6.22, the profiles of the cutting forces show good agreement between the experiments and the simulations. The difference between the predicted and experimental maximum cutting forces in the three presented cases was always less than 6%.

The prediction error of maximum and average cutting forces between the simulation and experimental data is shown in Table 6.3. The prediction errors using Bao's conventional micro-milling force model [6]) and Ding's vibration-assisted milling force model [1] are also listed for comparison. It can be noted that the conventional cutting force model is not suitable for the VAM, because the prediction errors are larger than 10% when using Set 1 parameters, and for the Set 2 and 3 parameters the errors are larger than 30%. Ding's model provides a cutting force model on the basis of the numerical simulation and has

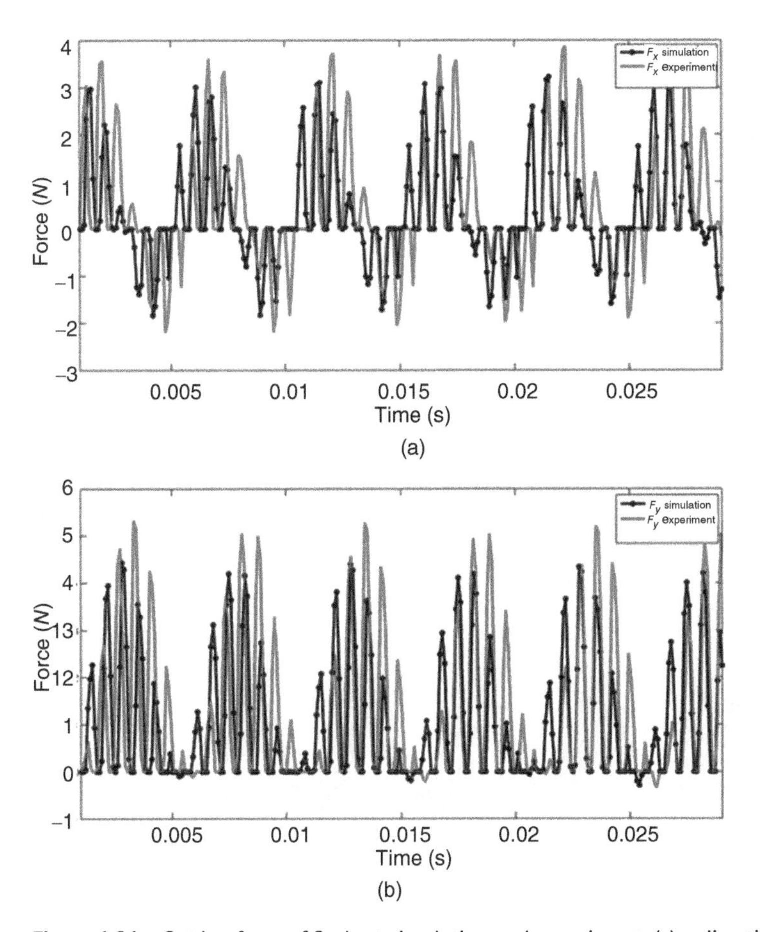

Figure 6.21 Cutting force of 2nd set simulation and experiment. (a) x direction; (b) y direction.

good prediction accuracy for Set 1 and 2 parameters, but for Set 3 it has a large prediction error (>10%). As can be seen from Table 6.3, the prediction of cutting force by using the model proposed in this study shows better accuracy than these previous models, especially for Set 3 with TWS type III.

As shown in Figure 6.23, three types of chips, namely "block type," "C type," and "continuous type" are found, during the VAMILL process. Moreover, it can be noted that the formation of each type of chip is related to the occurrence of each type of TWS.

In the machining experiment of the first set, no TWS occurred in one cutting path. Meanwhile, the "continuous type" chip was generated, thereby indicating that the formation process of continuous chip is not affected by TWS. On the other side, in the experiments of second and third sets, owing to the occurrence of TWS, "block type" and "C type" of the chips are found, which agree well with the results of the kinematic analysis. Therefore, the experimental results verify the correctness and reliability of the separation requirements obtained from the kinematic analysis.

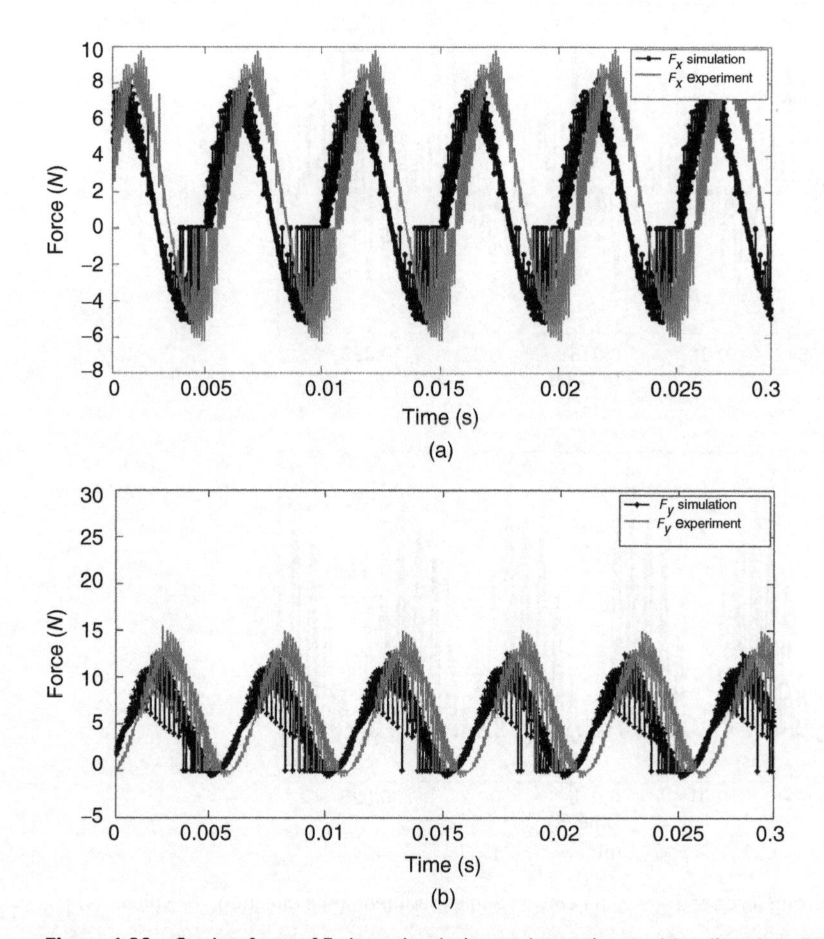

Figure 6.22 Cutting force of 3rd set simulation and experiment. (a) x direction; (b) y direction.

Table 6.3 The prediction error of maximum and average cutting force between the simulation and experimental data.

Cutting force	The proposed model in this paper			Ding's cutting force model			Conventional micro-milling cutting force model		
	Set 1 (%)	Set 2 (%)	Set 3 (%)	Set 1 (%)	Set 2 (%)	Set 3 (%)	Set 1 (%)	Set 2 (%)	Set 3 (%)
ΔF_x	4.34	4.33	6.72	4.52	4.83	10.53	10.23	30.2	35.6
ΔF_y	4.52	4.46	5.98	4.67	5.12	10.34	10.35	28.9	34.7
$\overline{F_x}$	3.55	3.42	4.73	3.82	4.12	8.34	8.23	25.4	30.2
$\overline{F_y}$	3.62	3.58	4.82	3.78	4.83	8.53	8.52	24.9	31.3

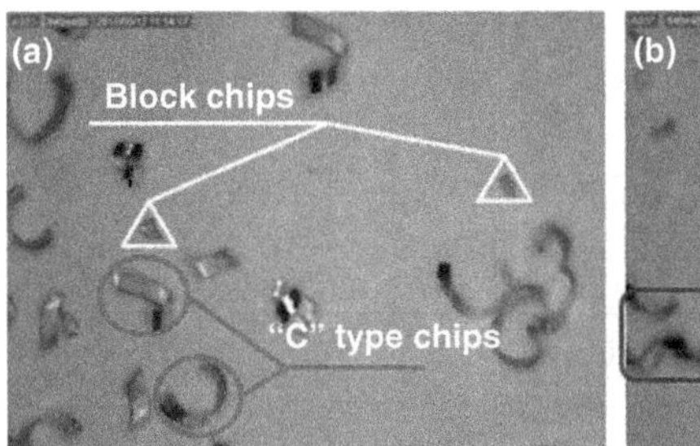

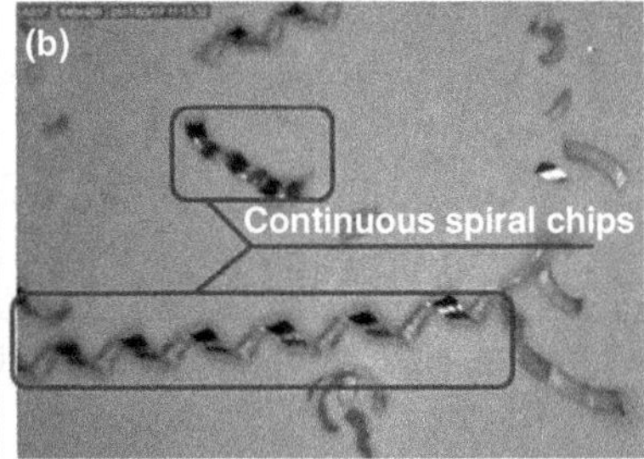

Figure 6.23 Micro-milling machine tool with vibration-assisted system; (a) chips with TWS; (b) chips with no TWS. Source: Courtesy of Dehong Huo, Wanqun Chen, and Lu Zheng.

6.4 Concluding Remarks

This chapter has provided a critical review of research into cutting forces in VAM. A mathematic cutting force model for VAM is developed taking into consideration the unique tool trajectory and cutting tool geometry. The conditions of TWS and its effect on cutting force are studied. FE modeling and machining experiments were considered to evaluate the mathematic cutting force model and the results of simulation and experimentation agree well.

References

1 Ding, H., Chen, S.J., and Cheng, K. (2010). Two-dimensional vibration-assisted micro end milling: cutting force modelling and machining process dynamics. *Proc. Inst. Mech. Eng. Part B J. Eng. Manuf.* 224: 1775–1783. https://doi.org/10.1243/09544054JEM1984.

2 Brehl, D.E. and Dow, T.A. (2008). Review of vibration-assisted machining. *Precis. Eng.* 32: 153–172. https://doi.org/10.1016/j.precisioneng.2007.08.003.

3 Zhang, C., Ehmann, K., and Li, Y. (2015). Analysis of cutting forces in the ultrasonic elliptical vibration-assisted micro-groove turning process. *Int. J. Adv. Manuf. Technol.* https://doi.org/10.1007/s00170-014-6628-3.

4 Zhang, C., Guo, P., Ehmann, K.F., and Li, Y. (2016). Effects of ultrasonic vibrations in micro-groove turning. *Ultrasonics* 67: 30–40. https://doi.org/10.1016/j.ultras.2015.12.016.

5 Zhang, X., Senthil Kumar, A., Rahman, M. et al. (2011). Experimental study on ultrasonic elliptical vibration cutting of hardened steel using PCD tools. *J. Mater. Process. Technol.* 211: 1701–1709. https://doi.org/10.1016/j.jmatprotec.2011.05.015.

6 Bao, W.Y. and Tansel, I.N. (2000). Modeling micro-end-milling operations. Part II: tool run-out. *Int. J. Mach. Tools Manuf* 40: 2175–2192. https://doi.org/10.1016/S0890-6955(00)00055-9.

7 Abootorabi Zarchi, M.M., Razfar, M.R., and Abdullah, A. (2012). Investigation of the effect of cutting speed and vibration amplitude on cutting forces in ultrasonic-assisted milling. *Proc. Inst. Mech. Eng. Part B J. Eng. Manuf.* 226: 1185–1191. https://doi.org/10.1177/0954405412439666.

8 Shen, X.H., Zhang, J.H., Li, H. et al. (2012). Ultrasonic vibration-assisted milling of aluminum alloy. *Int. J. Adv. Manuf. Technol.* 63: 41–49. https://doi.org/10.1007/s00170-011-3882-5.

9 Lin, J., Guan, L., Lu, M. et al. (2017). Modeling and analysis of the chip formation and transient cutting force during elliptical vibration cutting process. *AIP Adv.* 7 doi: 10.1063/1.5006303.

10 Cerniway, M.A. (2001). Elliptical diamond milling: kinematics, force, and tool wear. Dissertation, North Carolina State University. pp. 61–84.

7

Finite Element Modeling and Analysis of Vibration-Assisted Machining

7.1 Introduction

The machining of new materials used in aerospace, defense, aeronautical, medical, and electronics industries, possessing wear resistance, high strength at elevated temperature, and resistance to chemical degradation creates enormous challenges for the researchers. The main obstacle in the commercialization of these materials is their poor machinability. Excessive tool wear, difficulty in chip formation, poor surface quality, high temperature, and/or cutting forces are the bottlenecks of conventional machining. Moreover, the low thermal conductivity, high specific heat, and high strain hardening in conjunction with chemical reactivity with most of the cutting tool materials also result in extreme difficulty in machining such advanced engineering materials. Vibration-assisted machining presents significant advantages in improving the processability of difficult-to-machine materials, and therefore has broad application prospects.

In addition, the world is experiencing a growing demand for miniaturized products in various applications. Compared with other micro-manufacturing methods, micro-milling is believed to be the most flexible micromachining process with the capability to generate a wide variety of complex micro-components and microstructures due to its capability to process a wide range of materials, this enabling complex geometric features to be machined with a simple setup [1, 2]. Although micro-milling is promising for the manufacturing of miniaturized products, problems such as poor surface quality, edge burr generation, excessive tool wear, and short tool life make micro-milling of high-quality components more challenging.

During the machining process, the well-known size effect is associated with the tool edge radius, workpiece nonhomogeneity with respect to the tool/cut size, negative rake angles, and minimum chip thickness effects in the workpiece material and has been identified as a critical factor in determining the process performance [3]. The size effect has been reported to increase burr formation and exacerbate tool wear. The low ratio of undeformed chip thickness to tool edge radius leads to ploughing, poor edge quality, and burrs, and this impedes the attainment of a good finished surface and hinders functional compliance [4–6]. The post-processing of micro-components is extremely difficult and may cost more. The application of conventional deburring techniques may introduce dimensional errors and residual stresses in the components [7]. Therefore, the suppression of burr formation in micro-milling is of great significance.

Vibration Assisted Machining: Theory, Modelling and Applications,
First Edition. Lu Zheng, Wanqun Chen, and Dehong Huo.

Vibration-assisted machining is an external energy-assisted machining method in which vibration at high frequency and small amplitude is applied to the tool or workpiece to improve the material removal process. Various aspects of vibration-assisted methods with different cutting processes such as turning, drilling, and milling have been investigated by many researchers. A finite element model was developed by Amini et al. [8] to study the machining stress and force under the effect of various process parameters. They found that the amplitude of ultrasonic vibration of the cutting tool plays a significant role in determining the machining force. Nategh et al. [9] established a kinematics model describing the relative movement between the cutting tool and workpiece in vibration-assisted turning process. A toothed pattern was found on the lateral surface mainly in pressed and machined regions. In addition, the pressed region can only be observed when the cutting speed exceeds a critical value. Amini et al. [10] studied the effect of surface texture on the tribological properties in ultrasonic vibration-assisted face turning process. Microdimples on the surface were found to improve surface hardness and therefore increase the wear resistance and decrease friction coefficient. A study on the ultrasonic vibration-assisted drilling of Inconel 718 superalloy was conducted by Liao et al. [11]. A 2.7 times increase in drill life could be obtained with a small vibration amplitude. The ultrasonic vibration-assisted method also shows a positive effect in grinding of hard and brittle materials. Wang et al. [12] presented a mathematical model stating that grain track overlapping was the main factor contributing to the grinding force reduction and improvements in surface quality. Owing to the superiority of micro-milling in micro-manufacturing and the advantages of vibration assistance in machining process, it is worth investigating the strategies and conditions of vibration-assisted micro-milling for high-quality miniaturized components [13, 14]. Many researchers have conducted the investigation on vibration-assisted milling, which is mainly focused on surface texture generation and improvement in machining quality. For surface texture generation, Uhlmann et al. [15] found that the targeted high-frequency vibration of a workpiece influenced the surface topography and surface roughness of products in micro-milling. Tao et al. [16] realized the formation of surface texture by using feed-direction ultrasonic vibration-assisted milling. They found that a fish scale structure can be generated by vibration-assisted milling with optimal machining parameters. Chen et al. [17] investigated a new surface texture formation method in nonresonant vibration-assisted milling, where two specific types of surface textures – wave and fish scale are obtained by combining different machining parameters. Börner et al. [18] found that by using the defined cutting edge geometry and kinematics, the ultrasonic vibration-assisted milling represented a suitable method for a reproducible generation of a defined microstructure. For the machining quality improvement in vibration-assisted milling, Lian et al. [19] showed that inducing ultrasonic vibration during high-speed milling of small metal products can lead to smoother, higher quality surfaces. Shen et al. [20] investigated the influence of the vibration direction on the surface roughness and found that the vibration applied in feed direction was better than cross-feed direction. Ding et al. [21, 22] found that feed per tooth has a significant effect on the height of the top burr, and the use of vibration-assisted cutting in micro-end milling could minimize the size effect and improve the cutting performance, thereby reducing the height of the top burr. Richard et al. [23] studied the vibration-assisted milling for the machining of carbon fiber-reinforced polymers, and showed that through optimization of vibration characteristics such as form, frequency and amplitude, fraying behavior could be reduced. The control of burr formation in micro-milling was investigated by Chen et al. [7]. They found that due to the vibration assistance in the feed direction, up-milling and down-milling take place periodically on both cutting-in and cutting-out sides. This results in reduced burr formation.

Vibration-assisted machining has a huge potential advantage in micro-texture surface generation and high-quality surface finishing. However, most previous studies are based on the experimental analysis, and the cutting mechanism in vibration-assisted machining is still not clear, which limits the further expansion its application. In this chapter, the influence of vibration-assisted machining on the size effect and material removal mechanism is investigated using finite element modeling, and model validation was conducted by comparison of the results with experimental data.

7.2 Size Effect Mechanism in Vibration-Assisted Micro-milling

Figure 7.1 illustrates the material removal mechanisms in the micro-milling process. It can be noted that in the region that the cutter engages with the workpiece, the uncut chip thickness is smaller than a certain critical value h_c. Only elastic deformation occurs in the workpiece. The deformed material fully recovers to its original position without chip formation. As the cutting process advances, the chip is formed when the uncut chip thickness becomes larger than h_c. The value of h_c, which is known as minimum chip thicknesses, is found to be around 0.15–0.4 of the edge radius [24]. During the machining process, the cutting force, tool wear, surface roughness, and burr formation could be reduced by employing

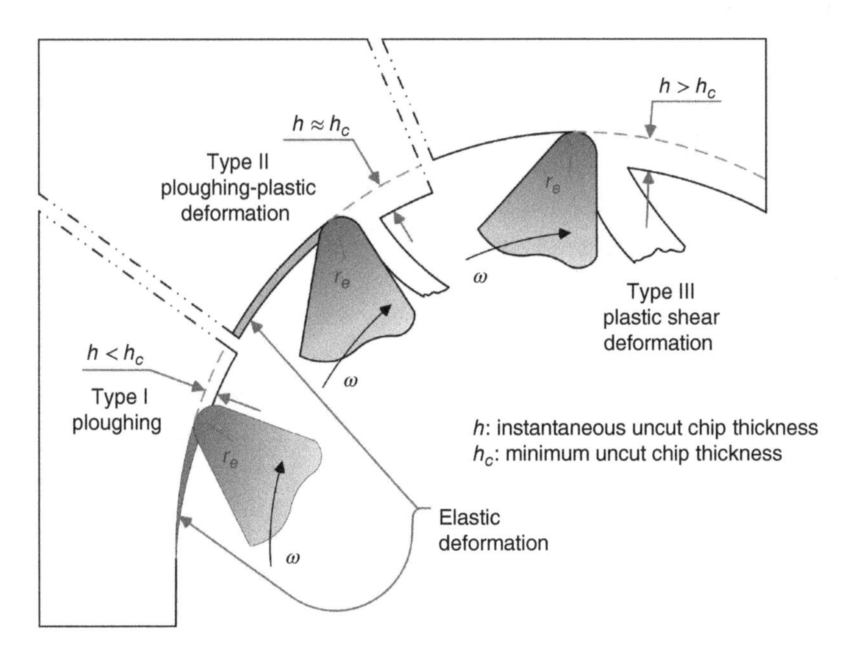

Figure 7.1 Material removal mechanisms in micro-milling.

the uncut chip thickness, which is larger than the minimum chip thickness, and the stability of the process is improved simultaneously. However, the minimum chip thickness cannot be avoided in the whole machining path given the micro-milling machining path, e.g. in the cutting-in region as shown in Figure 7.1.

By applying elliptical vibration in the micro-milling process, the instantaneous cutting thickness is significantly changed compared with the conventional micro-milling, especially in the cutting in and cutting out area. Therefore, the ploughing effect can be reduced.

The finite element cutting simulations in the ploughing region shown in Figure 7.1 are established for the conventional micro-milling and vibration-assisted milling processes. The machining parameters used in the simulation are listed as follows: the cutting speed is 2.1 m/s; the cutting depth is 0.1 μm; the cutting edge radius of the tool is 1 μm. The vibrations are applied to the workpiece in the feed and cross-feed directions with a frequency of 10 kHz and amplitude of 0.5 μm. The phase angle between the two vibration signals is $\pi/2$, which forms an elliptical vibration in the micro-milling process.

Figure 7.2a illustrates the variation in uncut chip thickness in the cutting path as shown in Figure 7.1. By assuming that h_c is around 0.25 of the edge radius, it can be found that in conventional micro-milling, the instantaneous cutting thickness increases from zero to the maximum uncut chip thickness with the rotation of the cutter, and after a specific time t, the instantaneous cutting thickness become larger than h_c, and chips start to form, as shown in Figure 7.2b with an obvious ploughing region. However, with the specified vibration parameters, it can be noted that instantaneous cutting thickness fluctuates severely, resulting in the area of the ploughing region being significantly reduced compared with that in conventional micro-milling.

7.2.1 FE Model Setup

A two-dimensional finite element model was developed to reveal the mechanism in vibration-assisted machining process using the commercial software Abaqus/Explicit v6.14-4. Arbitrary Lagrangian Eulerian (ALE) formulation is adopted to avoid excessive

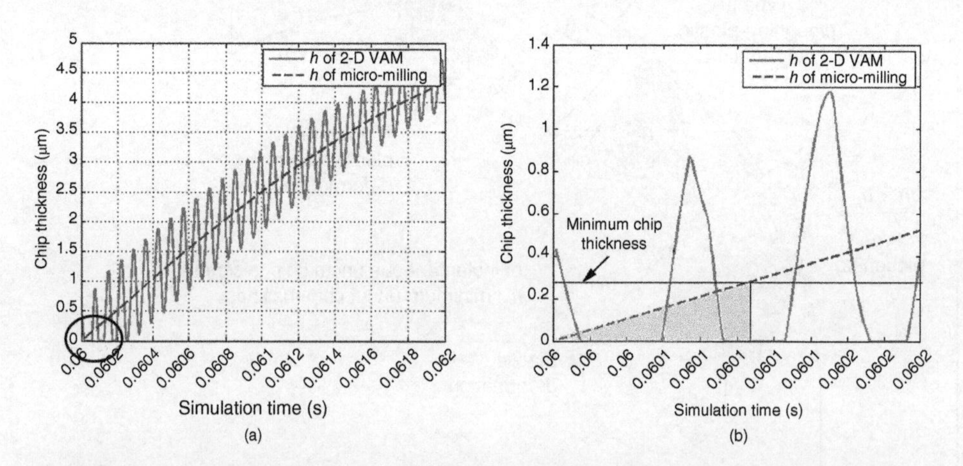

Figure 7.2 Comparison of uncut chip thickness between conventional and vibration-assisted micro-milling. (a) Uncut chip thickness, (b) enlarged figure.

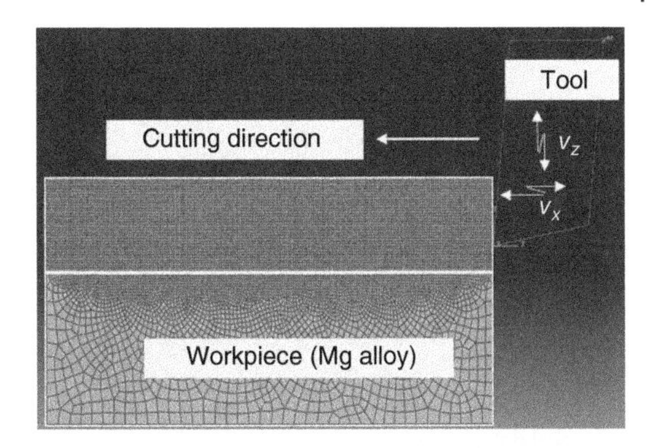

Figure 7.3 FE model of 2D vibration-assisted machining.

distorted element. The cutting tool was defined as an analytical rigid body with a predefined cutting speed and vibration amplitude and frequency. A schematic representation of the finite element (FE) model is shown in Figure 7.3.

The workpiece is treated as a deformable thermo-elastic-plastic material with a quadrilateral continuum element and fracture criteria. Johnson–Cook (JC) materials constitutive and failure models of material were used to describe the plastic deformation and failure mechanism in the machining process.

$$\overline{\sigma} = [A + B(\overline{\varepsilon}^{pl})^n]\left[1 + C\ln\left(\frac{\dot{\overline{\varepsilon}}^{pl}}{\dot{\varepsilon}_0}\right)\right]\left[1 - \left(\frac{T - T_{room}}{T_{melt} - T_{room}}\right)^m\right] \tag{7.1}$$

where $\overline{\sigma}$ is the flow stress, $\overline{\varepsilon}^{pl}$ is the plastic strain, $\dot{\overline{\varepsilon}}^{pl}$ is the plastic strain rate, $\dot{\varepsilon}_0$ is the reference strain rate, T is the workpiece temperature, and T_{melt} and T_{room} are the material melting and ambient temperature. Coefficient A is the yield strength, B is the hardening modulus, C is strain rate sensitivity coefficient, n is the hardening coefficient, and m is the thermal softening coefficient. The mechanical properties and material constants of workpiece in J–C model were listed in Table 7.1.

$$\overline{\varepsilon_f^{pl}} = (d_1 + d_2 e^{d_3\eta})\left[1 + d_4\ln\left(\frac{\dot{\overline{\varepsilon}}^{pl}}{\dot{\varepsilon}_0}\right)\right]\left[1 + d_5\left(\frac{T - T_{room}}{T_{melt} - T_{room}}\right)\right] \tag{7.2}$$

The free thermal-displacement quad-dominated meshing technique is used with a combination of advancing front algorithm for both matrix and particles. This model is meshed with 4-nodes quad-dominated element (CPE4R). The average mesh size is 5 μm. The surface-to-surface contact model was applied between the external surface of the tool and the node points of machining area. In this chapter, due to the high cutting speed and high friction caused by the size effect, the high temperature generated is believed to make a significant contribution to the plasticity behavior of the matrix material. This would make the sticking zone obvious during chip formation process. Therefore, Coulomb friction law combined with sticking–sliding theory is used to simulate friction stress.

$$\tau_{sticking} = \mu\sigma_n \quad \text{when} \quad \mu\sigma_n < \tau_{lim} \tag{7.3}$$

$$\tau_{sliding} = \tau_{lim} \quad \text{when} \quad \mu\sigma_n \geq \tau_{lim} \tag{7.4}$$

Table 7.1 Mechanical properties and materials constant in J–C model for magnesium alloy.

Properties	Values
Density (ton/mm^3)	1378×10^{-12}
Young's modulus (MPa)	39 820
Poisson's ratio	0.35
T_{melt} (K)	873
$T_{transition}$ (K)	293
Thermal expansion (K^{-1})	25×10^{-6}
Thermal specific heat (mJ/ton K)	914×10^{6}
Conductivity (mW/mm K)	156
A (MPa)	153
B (MPa)	291.8
N	0.1026
M	1.5
C	0.013
d_1	0.5
d_2	0.2895
d_3	3.719
d_4	0.013
d_5	1.5

Source: Based on Ulacia et al. [25].

where τ_{lim} is the limiting shear stress on the interface, σ_n is the normal stress distribution along the rake face, μ is the coefficient of friction, $\tau_{sticking}$ is the friction shear stress along the sticking region, and $\tau_{sliding}$ is the friction shear stress along the sliding region. A sticking region forms at the vicinity of the cutting tool and the equivalent shear stress (Eq. (7.3)) can be determined by the coefficient of friction μ and the normal stress distribution along the rake face σ_n. Once the shear stress at interface reaches a critical value τ_{lim}, sliding regime (Eq. (7.4)) governs the friction process. A constant friction coefficient of $\mu = 0.5$ is used in this simulation.

The machining parameters used in the simulation are shown as follows: the cutting speed is 2.1 m/s; the cutting depth is 50 μm; and the cutting edge radius is 1 μm. The workpiece is fixed with all degrees of freedom, and the vibrations in x and y directions are applied to the tool with the given frequency and amplitude, while the phase of the two vibration signal is set as $\pi/2$.

The relative motion of tool tip to workpiece in vibration-assisted cutting is

$$\begin{cases} x(t) = vt + A\cos(2\pi ft) \\ z(t) = B\sin(2\pi ft) \end{cases} \tag{7.5}$$

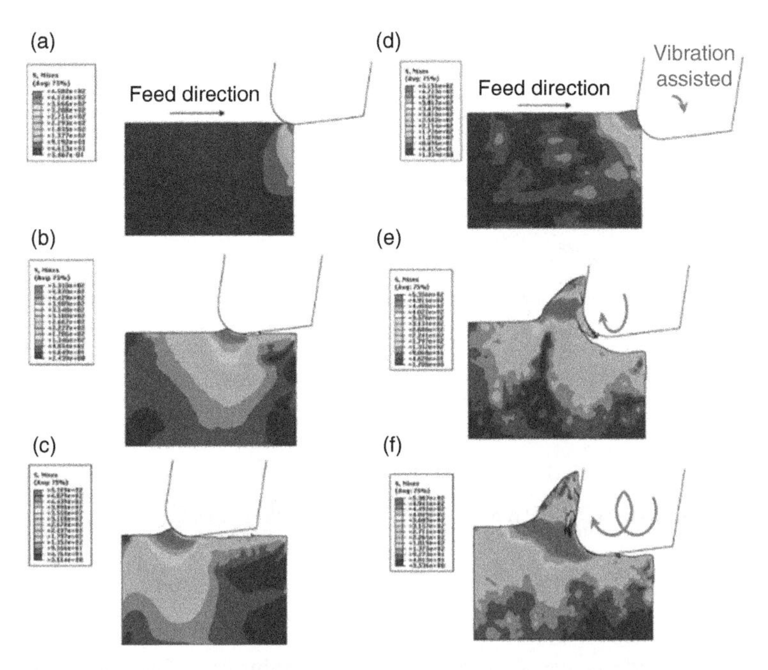

Figure 7.4 A comparison between conventional and vibration-assisted machining. (a–c) Conventional machining, (d–f) vibration-assisted machining.

where A and B are the vibration amplitudes in the x- and z-directions, respectively, f is the vibration frequency, and v is the cutting speed.

7.2.2 Simulation Study on Size Effect in Vibration-Assisted Machining

Figure 7.4a–c illustrates the simulated conventional machining process. It can be observed that as the cutting depth is smaller than the minimum chip thickness, the cutter squeezes the material forward during the cutting process with elastic deformation occurring in the workpiece. The deformed material fully recovers its original position after the tool pass without any chip formation. Figure 7.4d–f illustrates the simulated cutting process with elliptical vibration applied to the cutter. Within this process, the machining parameters are the same as in conventional machining, except for the assisted vibration. Based on the observation of Figure 7.4d, it can be found that due to the vibration applied to the cutter, the actual nominal instantaneous cutting thickness is greater than the minimum chip thickness although the initial cutting thickness is less than the minimum cutting thickness. Thus, the material directly comes into shearing removal mode instead of ploughing.

The specific cutting force is calculated by dividing the resultant cutting force by the section area of the cutting area. Observing Figure 7.5, the specific cutting force obtained from vibration-assisted cutting process (3.02 GPa) is smaller than that obtained from the non-vibration one (4.47 GPa). It can be said that when the vibration is added to the cutting process, a better capability to reduce the size effect at a small uncut chip thickness results, which is conducive to reducing tool wear rate.

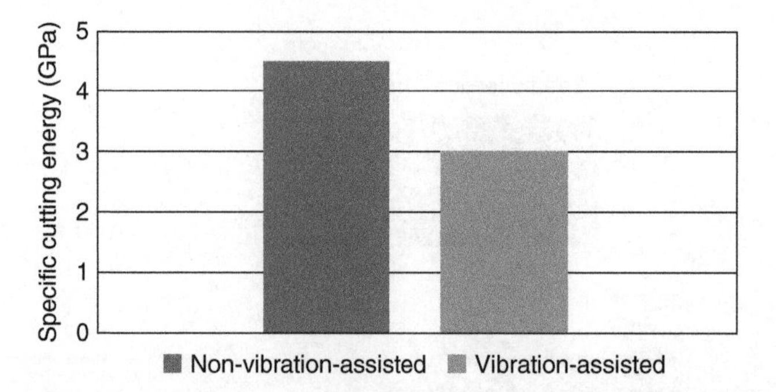

Figure 7.5 Specific cutting force obtained in conventional machining and vibration-assisted machining process.

7.3 Materials Removal Mechanism in Vibration-Assisted Machining

7.3.1 Shear Angle

As an ideal approach, many studies have been conducted to determine how the shear angle can be increased in the primary cutting zone. This increment causes some improvements in the cutting process such as reduction of deformed chip thickness, which finally results in the decrease of tool-chip contact length and power consumption [26]. Lotfi and Amini [27] studied the influence of the assisted vibration on the shear angle, as seen in Figure 7.6, where the effect of ultrasonic vibration on the shear angle and deformed chip thickness was studied. This was accomplished by comparing conventional turning (CT) with two methods of ultrasonic-assisted turning in the same step after running the simulations. The shear-angle graphs extracted from DEFORM 2D software plus the shear bands generated are shown in Figure 7.6a,b, respectively. Owing to excessive fluctuation in the values of shear angle, particularly in ultrasonic vibration-assisted turning (UVAT), the mean value of the graphs is graphically illustrated in Figure 7.6c in order to gain a better comparison. Accordingly, the shear angle increased when using linear vibration and there was a more significant increment when utilizing elliptical vibration. In 2D UVAT, this angle was taken a step forward in which it is higher than the ideal value. This caused the deformed chip thickness to be lower than its primary thickness. In this condition, the chip compression ratio is greater than one, followed by an increase in the deformed chip length. As seen in Figure 7.6d, the strain generated in the chip in 2D UVAT is approximately twice that of the values obtained in 1D UVAT and CT. These conditions can be effective on cutting forces and temperature distribution, as discussed in the following of this chapter. Furthermore, the experimental chips obtained during CT, 1D, and 2D UVAT are shown in Figure 7.7. The repeatability in measurement of shear angle is very low. However, the increment of this angle is clearly seen in UVAT methods compared to CT.

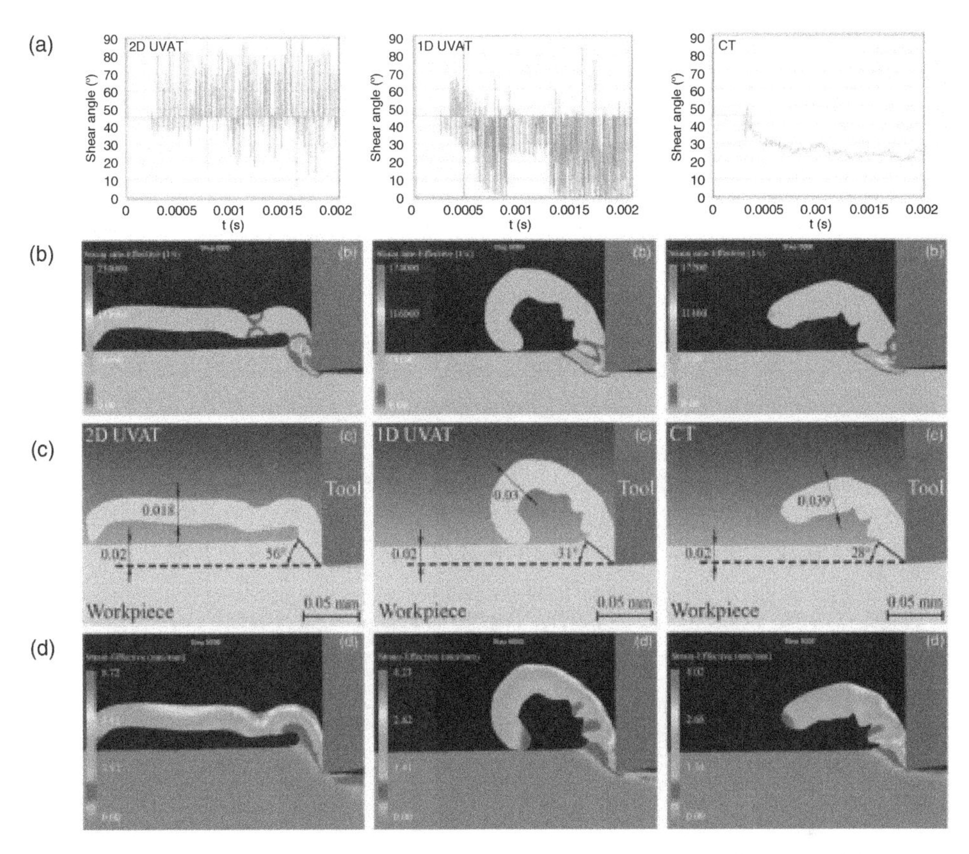

Figure 7.6 (a) Shear-angle graph, (b) strain rate and shear band, (c) chip thickness and the mean value of shear angle, and (d) strain in the deformed chip during CT, 1D, and UVAT process. Source: Lotfi and Amini [27].

Figure 7.7 Experimental deformed chips during (c) CT, (b) 1D, and (a) 2D UVAT process. Source: Courtesy of Dehong Huo, Wanqun Chen, and Lu Zheng.

7.3.2 Simulation Study on Chip Formation in Vibration-Assisted Machining

The simulated chip formation process occurring with plastically deformation at different elliptical vibration frequency is illustrated in Figures 7.8–7.10. The same FE model setup applied in Section 7.2.1 was used. As shown in Figure 7.1, during the cutting process advancement, the cutter enters the plastic shear deformation region. With the aim of observing the phenomena occurring in the cutting process more clearly, the machining

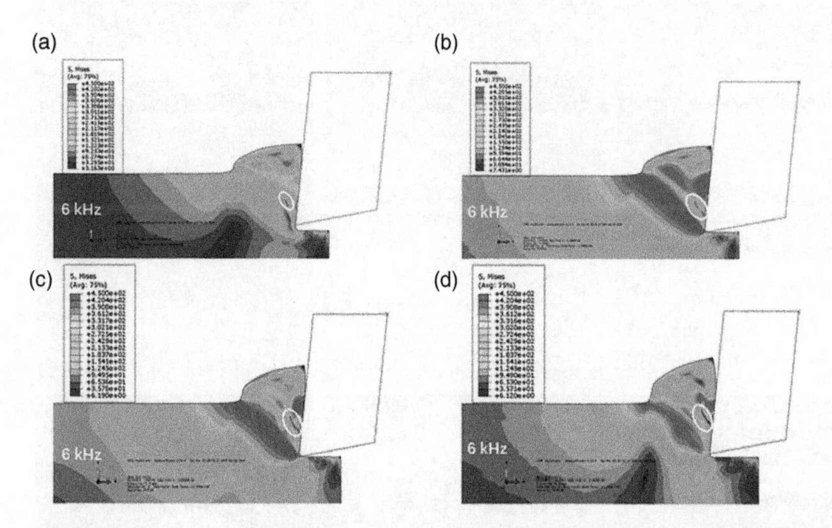

Figure 7.8 The chip formation with elliptical vibration of 6 kHz. (a–d) Cracks state at different times.

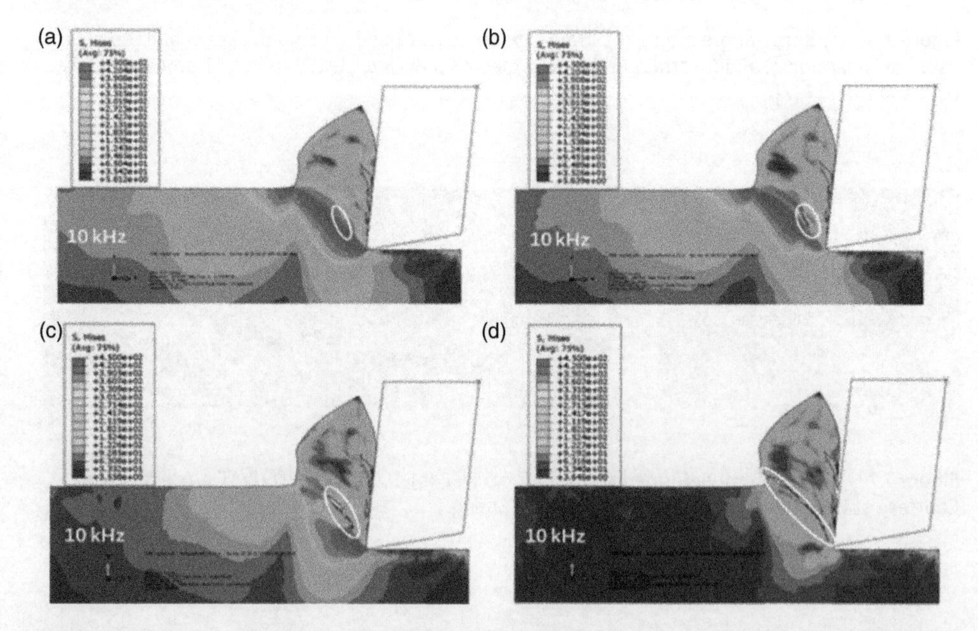

Figure 7.9 The chip formation with elliptical vibration of 10 kHz. (a–d) Cracks state at different times.

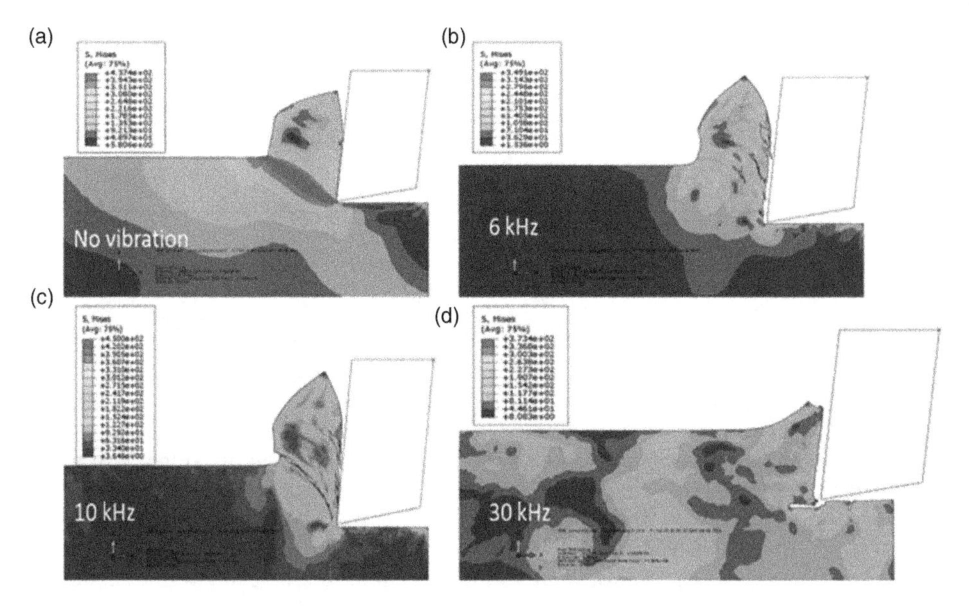

Figure 7.10 The chip formation with different vibration frequencies. (a–d) Cracks state at different times.

process is simplified to be orthogonal cutting. The machining parameters used in the simulation are shown as follows: the cutting speed is 2.1 m/s; cutting depth is 5 μm; the cutting edge radius is 1 μm.

Figure 7.8 illustrates the elliptical vibration with a frequency of 6 kHz. Cracks can be found on the chip root at tool–workpiece contact zone. Figure 7.8a shows a crack initiation on the chips, and with the tool continuously moving, another crack is generated on the chips. Then, the crack starts to grow and expand under the impact and extrusion of the cutter, as shown in Figure 7.8c,d. Finally, the cracks are located close to the rake face of the tool, which can be considered as the main reason for reducing the rake face wear and cutting force.

When the elliptical vibration frequency increases to 10 kHz, it can be noted that cracks are generated along the shear angle. Figure 7.9a,b shows that an initial crack is found in the shear plane. Then the cracks growth is initiated under the vibration impact of the cutting tool as shown in Figure 7.9c. Finally, the chips fracture along the shear angle resulting in block chips as shown in Figure 7.9d.

The chips formation process is compared at different vibration frequencies, as shown in Figure 7.10. It can be noted that the continuous chip is formed during the cutting process without vibration, and no obvious crack can be found during the machining process. When the elliptical vibration frequency increases from 6 to 30 kHz, the crack generated changes in direction from nearly at the rake face to the shear plane and finally along the upfeed direction, and becomes more pronounced. This phenomenon can be explained by the variation in tool trajectory at different vibration frequencies. As shown in Figure 7.11, with the increase in the vibration frequency applied in the machining process, the impact angle between the tool tip and the workpiece decreases ($\theta_1 > \theta_2 > \theta_3$). Thus, the impact direction of the vibration in the workpiece tends to horizontal with the vibration frequency

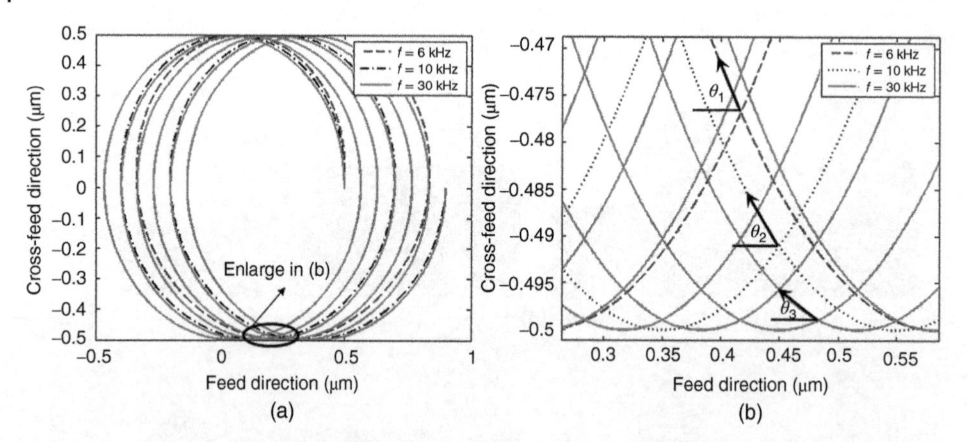

Figure 7.11 Comparison of tool trajectory with different vibration frequencies. (a) Tool trajectory, (b) enlarged figure.

increasing, and the cracks in the workpiece will occur along the impact direction, since the impact stress reaches its maximum value in this direction. As the changes in direction of the cracks are generated, chip morphology gradually changes from continuous to broken chips.

7.3.3 Characteristics of Simulated Cutting Force and von-Mises Stress in Vibration-Assisted Micro-milling

All FE simulations were performed for VAMM and CMM processes with identical cutting parameters. The intermittent nature of contact due to tool–workpiece separation in VAMM is the main reason for the differences in cutting stresses and forces in the two machining processes.

From the point of view of energy, the destruction of material can be considered to be a result of the impact stress exceeding the critical stress of the material. In the vibration-assisted cutting process, the material naturally moves faster and it is easier to be destroyed with the increase in tool vibration speed. The material destruction mechanism can be considered to be completely instantaneous under the action of high-speed impact loads, and the small impact per tooth will result in the plastic failure of the material.

Figure 7.12 demonstrates an example of stress distribution and cutting force signatures for VAMM and CMM processes. The simulation was performed for AISI 1045 steel, and the frequency and amplitude of the vibration in VAM were 20 kHz and 5 μm, respectively. Apparently, in VAMM the maximum stress is larger than that in CMM. Figure 7.13 shows that the cutting stress at point A changes with cutting time, and it can be noted that in VAMM levels of stress clearly fluctuate. At time t, the stress of point A is larger than its fracture stress even though the tool has not been cut to this point; thus, it means that the crack is generated before the cutting point, which results in the decrease of the cutting force.

In terms of the effect of vibration added to the characteristics of cutting force, it can be seen from Figure 7.14 that the cutting force of conventional cutting is about 3 N, while a cyclical fluctuation of cutting force is found when the vibration with a frequency of 6 kHz is applied. The maximum cutting force is equivalent to the cutting force of ordinary

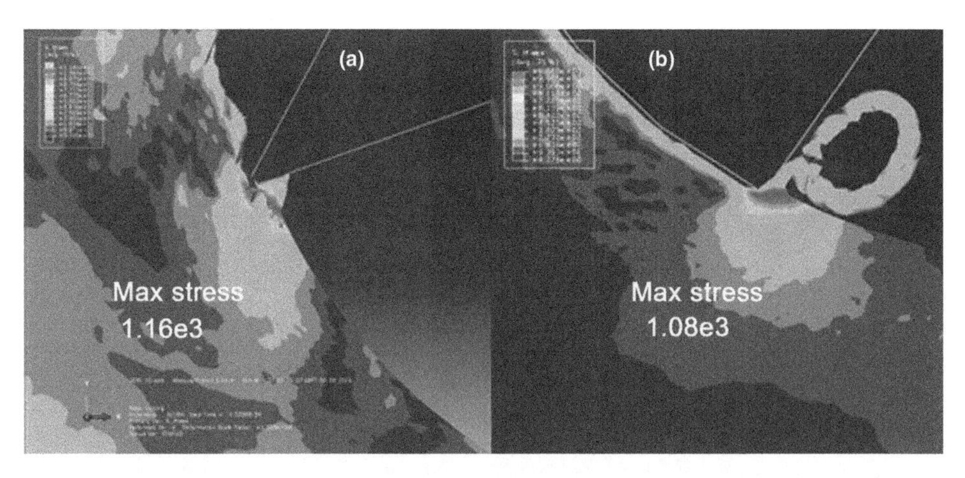

Figure 7.12 Stress distribution of (a) VAMM and (b) CMM.

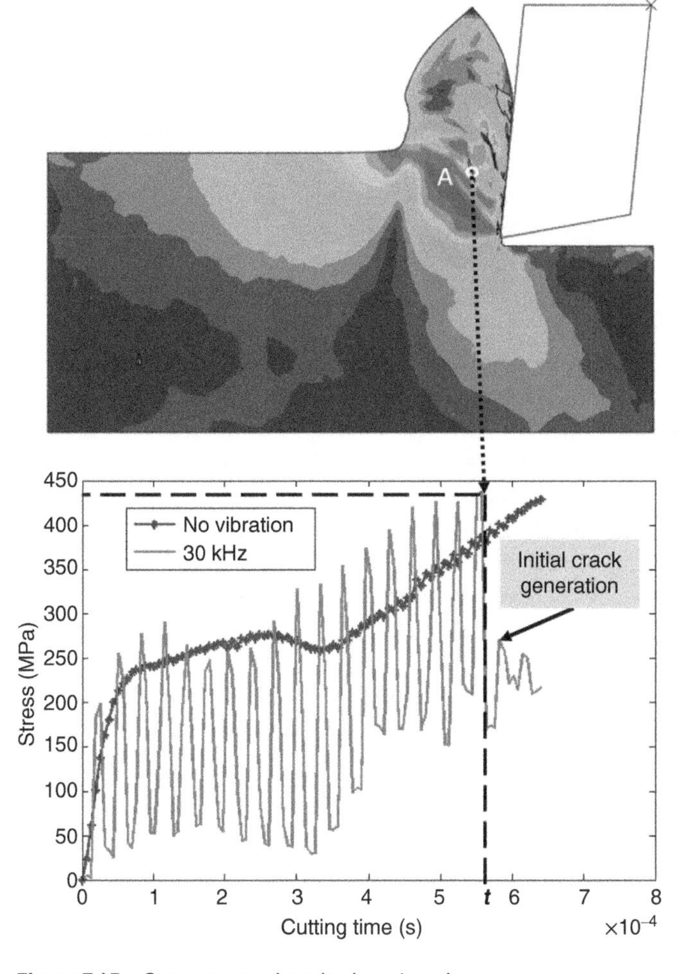

Figure 7.13 Stress comparison in time domain.

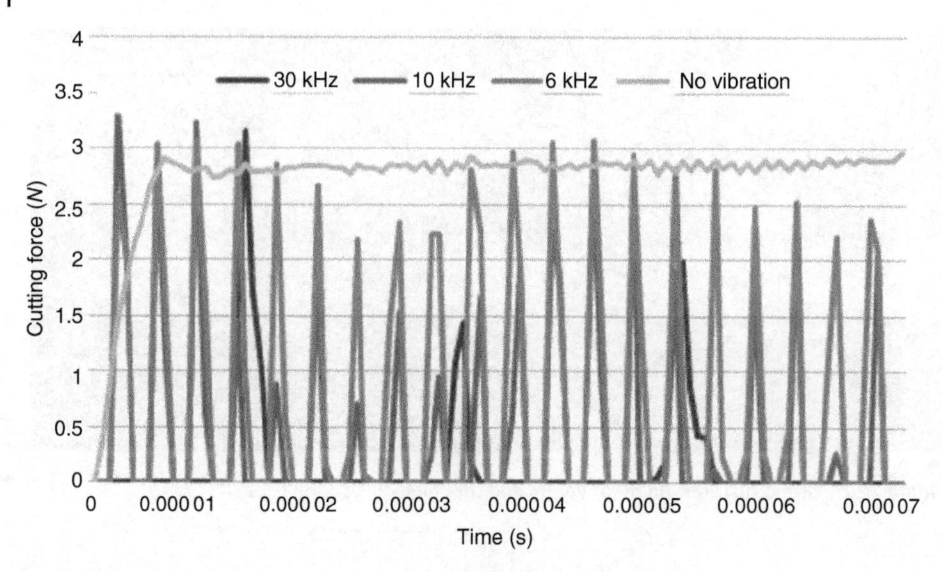

Figure 7.14 Cutting force comparison with different vibration frequencies.

cutting, but the average cutting force is significantly lower than the average cutting force obtained from ordinary cutting. While the cutting process continues, the crack is continuously formed on the chip in contact with the rake face. Therefore, the peak cutting force is sometimes lower than that in ordinary cutting; when the vibration frequency is increased to 10 kHz, the cutting force shows periodic fluctuations, but its amplitude fluctuates greatly. When the tool suddenly comes into contact with the workpiece, the peak cutting force is found to be approximately 3.5 N, which is greater than the cutting force in ordinary cutting. This can be attributed to the impact of the tool when it is in contact with the workpiece. After that, as described above, the crack is continuously formed on the chip in the shear plane, and therefore, the peak cutting force in vibration cutting is reduced to nearly 2 N; this is significantly lower than the ordinary cutting force. When the vibration frequency is further increased to 60 kHz, the cracking of the material caused by the vibration of the cutting tool becomes more obvious. The effective cutting force interval becomes larger. Thus the average cutting force is decreased to approximately 0.16 N.

It can be noted that for the steel machining the cutting force frequency will be increased with the vibration frequency; however, for a soft material such as magnesium alloy whose damage stress is much smaller than stainless steel, cracks are easy to generate in the vibration-assisted machining process. When the frequency of vibration exceeds a certain value, the fluctuation frequency of the cutting force will be inconsistent with the vibration frequency.

7.4 Burr Control in Vibration-Assisted Milling

As mentioned earlier, due to the small feature size in the micro-milling process, post-process deburring is particularly difficult or even impossible to perform. Moreover, the deburring process would always affect dimensional accuracy, and may introduce additional residual stress and damage on the workpiece. Therefore, the minimization and control of burr becomes more important.

Currently, the development of effective burr control techniques in micro-milling remains a pressing task. In order to suppress and control burr formation, numerous studies have been conducted on the mechanism by which burrs are formed and on the optimization of machining parameters. Lekkala et al. [28] characterized and modeled burr formation in micro-end milling. The influence of the main process parameters of speed, feed rate, depth of cut, tool diameter, and the number of flutes on the formation of various burr types was studied. It was found that the burr height tends to decrease with increasing feed rate. Biermann and Steiner [29] studied micro-burr formation in the milling process of austenitic stainless steel. They pointed out that in the up-milling process of a slot, the top burr height is evidently reduced, but it tends to increase with cutting speed. By conducting finite element simulation, Yadav et al. [30] investigated the micro-milling of Ti6Al4V and found that burrs are usually formed on the up-milling side at the exit of the micro-milling tool. Saptaji et al. [31] studied the effect of side edge angle and rake angle on top burrs in micro-milling and found that as the taper angle increases, the top burrs were reduced significantly. Kou et al. [32] proposed a burr reduction method with a layer of supporting material on the workpiece. Most burrs were formed in the sacrificial supporting material and a high-quality machined surface was obtained after removing the supporting material. Kumar et al. [33] reviewed recent advances in characterization, modeling, and control of burr formation in micro-milling. They concluded that optimization of cutting parameters, coating, hybrid cooling/lubrication, and the supporting material method could be used to minimize or prevent burr formation. However, the current methods controlling burr formation in micro-milling are either very complicated or of limited effectiveness. The following sections state the kinematic analysis and finite element simulation on burr formation process.

7.4.1 Kinematics Analysis

The kinematics of conventional micro-milling are shown in Figure 7.15a. For slot milling, one side of the slot is generated by up-milling, and the other by down-milling. With the cutter penetrating into the workpiece from the up-milling side, the workpiece material is squeezed and pushed. With the tool rotation and feeding of the workpiece, the uncut chip thickness increases. This results in an increase in the supporting effect on the uncut material and material removal due to shear action. With the continuous rotation of the cutter, workpiece material is removed by the cutting edge from the down-milling side, and the supporting effect of uncut material is smaller compared with the up-milling side. The uncut material is pushed out of the top of the slot, which is in the direction with the lowest resistance, thereby generating a large irregular tearing burr on the down-milling side. Therefore, for ductile material micro-milling, the surface quality on the up-milling side is better than that of the down-milling side as reported in [29, 30] and confirmed in this research.

Based on the top burr formation mechanism, a burr control strategy with the aid of vibration is proposed. As depicted in Figure 7.15a, when the workpiece is moved along the tool rotation direction in the conventional micro-milling process, up-milling occurs. For

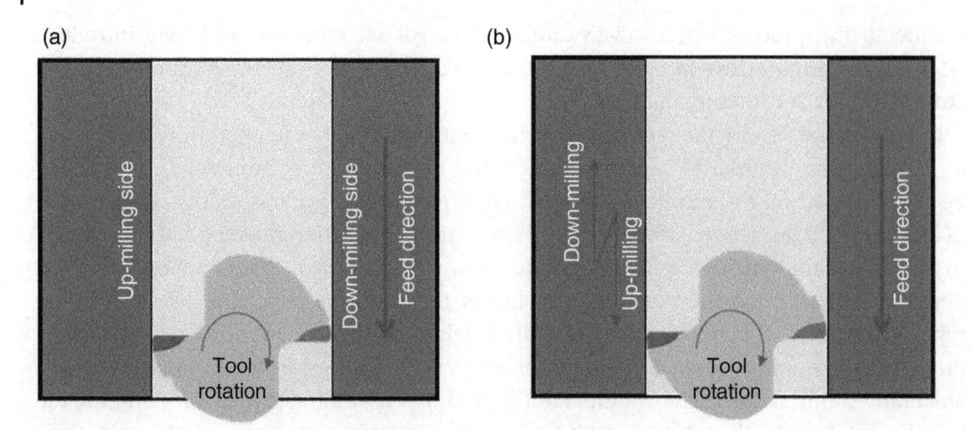

Figure 7.15 Schematic diagrams of slot milling: (a) conventional micro-milling, (b) vibration-assisted micro-milling.

a workpiece fed in the opposite direction, down-milling occurs. Therefore, there is always an up-milling side and a down-milling side in slot milling, and the latter has a relatively poor edge quality. In this study considering the kinematics of the micro-milling process, by applying the appropriate vibration in feed direction up-milling and down-milling can alternately occur on both sides, as shown in Figure 7.15b. Moreover, when the workpiece is locally subjected to the impact load from the vibration-driven tool, deformation and fracture of workpiece material in the cutting zone will reach a maximum in a shorter time, which changes the material removal process. Material removal in the conventional milling process is dominated by the shearing action of the tool, while in vibration-assisted milling the material removal mechanism is dependent on the hybrid interactions of the impact and shearing. Thus, the formation of burrs is effectively suppressed due to the application of vibration.

7.4.2 Finite Element Simulation

Burr formation in the micro-milling process was simulated using an FE model (Figure 7.16). Titanium alloy (Ti6Al4V) was selected as the workpiece material. The dimensions of the workpiece in the FE model are $2 \times 1 \times 0.2\,\text{mm}^{-3}$ and the minimum element size is set at 10 μm. A 1 mm diameter tungsten carbide end mill with a 3 μm cutting edge radius is used in the simulation. The machining parameters used are tool rotational speed $n = 40\,000$ rpm, feed per tooth $f_z = 15$ μm, and axial depth of cut ap = 50 μm. Vibration (frequency 5000 Hz, amplitude 10 μm) is applied to the workpiece in the feed direction. The Johnson–Cook (JC) model of material constitutive and damage models are used to model material plasticity and damage respectively. The coefficient of friction between tool and workpiece is set at 0.6.

Figure 7.17 shows the simulation results for conventional micro-milling and vibration-assisted micro-milling, respectively. It can be seen from Figure 7.17a that the burr size in down-milling side is larger than that in the up-milling side in conventional micro-milling. However, on the slot edges machined by vibration-assisted milling, the burrs in the down-milling side are reduced significantly.

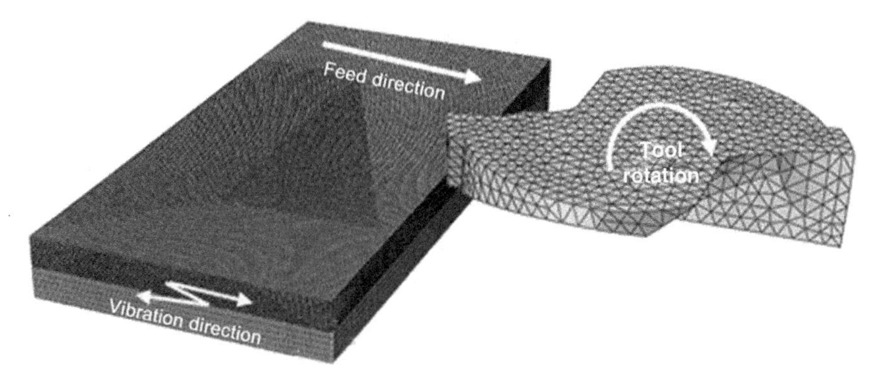

Figure 7.16 Finite element model of slot milling with vibration assistance in feed direction.

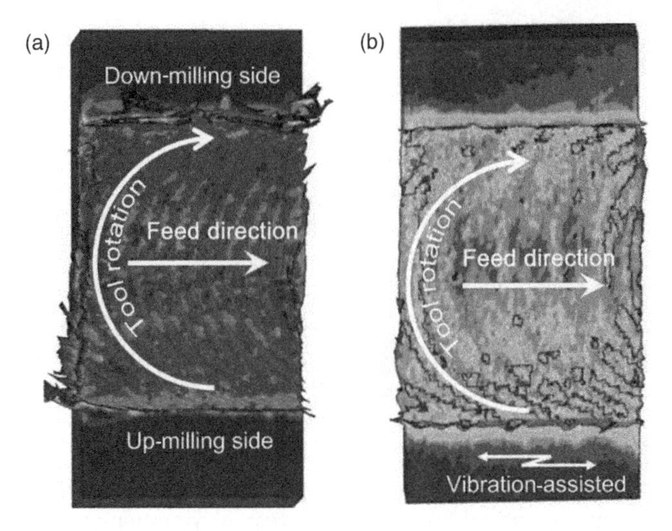

Figure 7.17 Slot micro-milling simulation results: (a) conventional, (b) vibration-assisted.

7.5 Verification of Simulation Models

Machining experiments were performed to investigate the effect of added vibration on tool wear rate and model validation was conducted in terms of chip and burr formation. Micro-milling experiments were performed on a Nanowave MTS5R micro-milling machine equipped with a high-speed spindle (maximum speed 80 000 rpm). The 2D vibration stage is installed on the Z axis guideway, as shown in Figure 7.18. The vibration stage is capable of imposing independent vibrations in the X and Y directions. The vibration stage design is based on two layers of flexure hinges, which minimizes the displacement couple effect in the two movement directions. The stage can realize 10 kHz vibration with a maximum amplitude of 20 μm and a low-frequency coupling less than 0.3 μm. Uncoated tungsten carbide end milling tools of 1 mm diameter with two cutting flutes are used in the experiment. The edge radius, helix angle, and diameter of the milling tool are 3 μm, 30°, and 1 mm respectively.

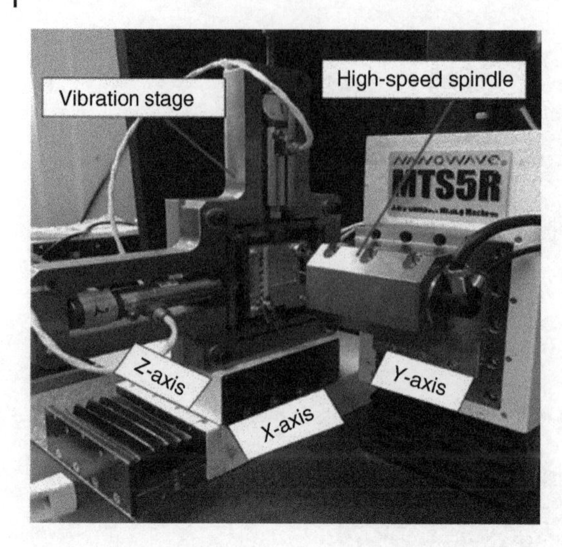

Figure 7.18 Ti6Al4V machining experiments setup. Source: Chen et al. [7]. © 2018, Elsevier.

7.5.1 Tool Wear and Chip Formation

Magnesium alloy was used to experimentally investigate tool wear and chip formation. The comparison of conventional and vibration-assisted micro-milling was conducted with the following machining parameters: cutting depth is 50 μm, spindle speed is 40 000 rpm, feed per tooth is 0.1 μm, which obviously fall into the plough region. The vibrations are applied in the feed and cross-feed directions on the workpiece with the frequencies of 10 kHz and amplitude of 0.3 μm. The phase angle between the two vibration signals is $\pi/2$, which results in an elliptical vibration in the micro-milling process. After cutting 20 mm × 20 mm long slots of magnesium alloy, it is found that the tool wear rate obtained from vibration-assisted micro-milling is significantly reduced when compared to that obtained from conventional micro-milling. The tool wear of vibration-assisted micro-milling mainly occurs at tool tip area, while for conventional micro-milling, the most worn area is tool tip and extends to the flank face of the tool. The average wear progression for both conventional and vibration-assisted machining was measured, as shown in Figure 7.19. The results show that the flank wear is reduced by almost 73% when vibration is added and drops from nearly 15 μm to less than 4 μm.

Figure 7.20 illustrates the generated chips with conventional micro-milling and vibration-assisted milling with 6 and 10 kHz, respectively. The vibration amplitude is 2 μm and the machining parameters are as follows: the cutting depth is 50 μm, spindle speed is 40 000 rpm, and feed per tooth is 5 μm. It can be found that the chip generated from conventional micro-milling is smooth and continuous. Significant cracks on the chips could be clearly observed after a vibration frequency of 6 kHz was applied. When the vibration frequency of 10 kHz was applied, block chips appeared, and this is in good agreement with the simulation results. The correctness of the simulation analysis is verified.

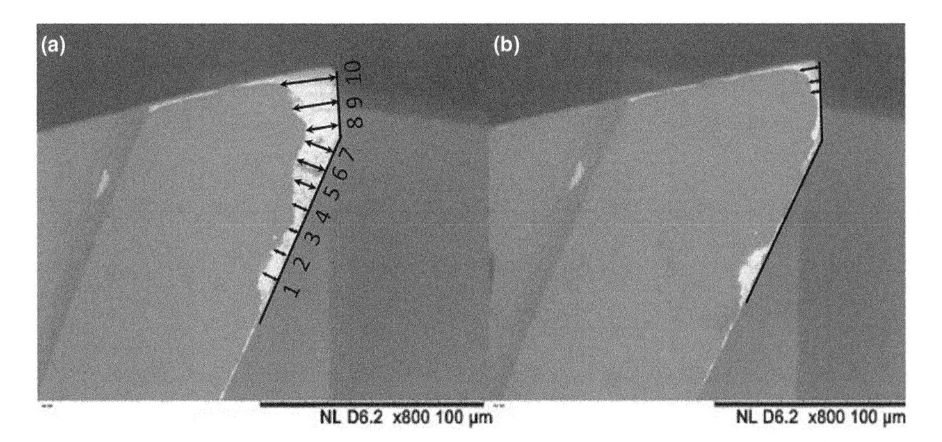

Figure 7.19 Tool wear and machined surface test results. (a, b) Tool wear in conventional and vibration-assisted micro-milling, respectively. Source: Chen et al. [34]. © 2019, Springer Nature.

Figure 7.20 Chips with different vibration frequencies. Source: Chen et al. [34]. © 2019, Springer Nature.

7.5.2 Burr Formation

The morphology of burrs obtained from conventional micro-milling and vibration-assisted micro-milling experiments were compared. Titanium alloy was selected as workpiece material. The simulation results indicate that the vibration and machining parameters used in the finite element cutting simulation were appropriate to reduce the top burr on the machined surface, so the same vibration and machining parameters are adopted in this experimental work. The work vibration in the Y direction (the feed direction) was suppressed.

As shown in Figure 7.21a, in conventional micromachining burrs of different size are formed on each side of the slot. It may be seen that more and larger burrs are formed on the down-milling side. The average top burr height on the down-milling side is around 15 μm, compared with 2 μm on the up-milling side. For the slots machined using vibration-assisted micro-milling, the burr height on the down-milling side is significantly reduced as shown in Figure 7.21b. The average top burr height on both the down- and up-milling sides is

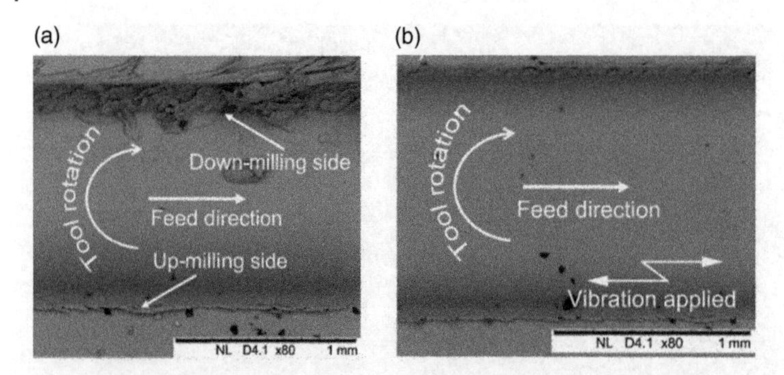

Figure 7.21 Burr generation. (a) Conventional assisted micro-milling; (b) vibration-assisted micro-milling. Source: Chen et al. [7]. © 2018, Elsevier.

around 2 μm. Compared with conventional micro-milling, the average top burr height on the down-milling side has been reduced by 87%.

7.6 Concluding Remarks

This chapter has investigated the cutting mechanism and chips formation process in vibration-assisted machining. Finite element simulation and machining experiments were conducted and compared with conventional micro-milling; more cracks were generated in the cutting region with vibration-assisted milling, resulting in decreased cutting force.

References

1 Huo, D., Lin, C., and Dalgarno, K. (2014). An experimental investigation on micro machining of fine-grained graphite. *Int. J. Adv. Manuf. Technol.* 72: 943–953.

2 Huo, D., Choong, Z., Shi, Y. et al. (2016). Diamond micro-milling of lithium niobate for sensing applications. *J. Micromech. Microeng.* 26: 095005.

3 Aramcharoen, A. and Mativenga, P.T. (2009). Size effect and tool geometry in micromilling of tool steel. *Precis. Eng.* 33 (4): 402–407.

4 Bissacco, G., Hansen, H.N., and Slunsky, J. (2008). Modelling the cutting edge radius size effect for force prediction in micro milling. *CIRP Ann. – Manuf. Technol.* 57 (1): 113–116.

5 De Oliveira, F.B., Rodrigues, A.R., Coelho, R.T., and De Souza, A.F. (2015). Size effect and minimum chip thickness in micromilling. *Int. J. Mach. Tools Manuf.* 89: 39–54.

6 Mian, A.J., Driver, N., and Mativenga, P.T. (2011). Identification of factors that dominate size effect in micro-machining. *Int. J. Mach. Tools Manuf.* 51 (5): 383–394.

7 Chen, W., Teng, X., Zheng, L. et al. (2018). Burr reduction mechanism in vibration-assisted micro milling. *Manuf. Lett.* 16: 6–9.

8 Amini, S., Soleimanimehr, H., Nategh, M. et al. (2008). FEM analysis of ultrasonic-vibration-assisted turning and the vibratory tool. *J. Mater. Process. Technol.* 201: 43–47.

9 Nategh, M., Razavi, H., and Abdullah, A. (2012). Analytical modeling and experimental investigation of ultrasonic-vibration assisted oblique turning, part I: kinematics analysis. *Int. J. Mech. Sci.* 63: 1–11.
10 Amini, S., Hosseinabadi, H.N., and Sajjady, S.A. (2016). Experimental study on effect of micro textured surfaces generated by ultrasonic vibration assisted face turning on friction and wear performance. *Appl. Surf. Sci.* 390: 633–648.
11 Liao, Y.S., Chen, Y.C., and Lin, H.M. (2007). Feasibility study of the ultrasonic vibration assisted drilling of inconel superalloy. *Int. J. Mach. Tools Manuf.* 47 (12–13): 1988–1996.
12 Wang, Y., Lin, B., Wang, S., and Cao, X. (2014). Study on the system matching of ultrasonic vibration assisted grinding for hard and brittle materials processing. *Int. J. Mach. Tools Manuf.* 77: 66–73.
13 Shen, X.-H., Zhang, J.H., Li, H. et al. (2012). Ultrasonic vibration-assisted milling of aluminum alloy. *Int. J. Adv. Manuf. Technol.* 63 (1–4): 41–49.
14 Chen, W., Huo, D., Hale, J., and Ding, H. (2018). Kinematics and tool-workpiece separation analysis of vibration assisted milling. *Int. J. Mech. Sci.* 136: 169–178.
15 Uhlmann, E., Perfilov, I., and Oberschmidt, D. (2015). Two-axis vibration system for targeted influencing of micro-milling. *Euspen's 15th International Conference & Exhibition* (05 June 2015). Leuven, Belgium.
16 Tao, G., Ma, C., Bai, L. et al. (2017). Feed-direction ultrasonic vibration-assisted milling surface texture formation. *Mater. Manuf. Processes* 32 (2): 193–198.
17 Chen, W., Zheng, L., Huo, D., and Chen, Y. (2018). Surface texture formation by non-resonant vibration assisted micro milling. *J. Micromech. Microeng.* 28 (2): 025006.
18 Börner, R., Winkler, S., Junge, T. et al. (2018). Generation of functional surfaces by using a simulation tool for surface prediction and micro structuring of cold-working steel with ultrasonic vibration assisted face milling. *J. Mater. Process. Technol.* 255: 749–759.
19 Lian, H., Guo, Z., Huang, Z. et al. (2013). Experimental research of Al6061 on ultrasonic vibration assisted micro-milling. *Procedia CIRP* 6: 561–564.
20 Shen, X.-H., Zhang, J., Xing, D.X., and Zhao, Y. (2012). A study of surface roughness variation in ultrasonic vibration-assisted milling. *Int. J. Adv. Manuf. Technol.* 58 (5–8): 553–561.
21 Ding, H., Chen, S.J., Ibrahim, R. et al. (2011). Investigation of the size effect on burr formation in two-dimensional vibration-assisted micro end milling. *Proc. Inst. Mech. Eng. Part B* 225 (11): 2032–2039.
22 Ding, H., Ibrahim, R., Cheng, K., and Chen, S.J. (2010). Experimental study on machinability improvement of hardened tool steel using two dimensional vibration-assisted micro-end-milling. *Int. J. Mach. Tools Manuf.* 50 (12): 1115–1118.
23 Zemann, R., Kain, L., and Bleicher, F. (2014). Vibration assisted machining of carbon fibre reinforced polymers. *Procedia Eng.* 69: 536–543.
24 Cheng, K. and Huo, D. (2013). *Micro-Cutting: Fundamentals and Applications*. Wiley.
25 Ulacia, I., Salisbury, C.P., Hurtado, I., and Worswick, M.J. (2011). Tensile characterization and constitutive modeling of AZ31B magnesium alloy sheet over wide range of strain rates and temperatures. *J. Mater. Process. Technol.* 211: 830–839.
26 Boothroyd, G. (1988). *Fundamentals of Metal Machining and Machine Tools*, vol. 28. New York: CRC Press.

27 Lotfi, M. and Amini, S. (2018). FE simulation of linear and elliptical ultrasonic vibrations in turning of Inconel 718. *Proc. Inst. Mech. Eng. Part E* 232 (4): 438–448.

28 Lekkala, R., Bajpai, V., Singh, R.K., and Joshi, S.S. (2011). Characterization and modeling of burr formation in micro-end milling. *Precis. Eng.* 35 (4): 625–637.

29 Biermann, D. and Steiner, M. (2012). Analysis of micro burr formation in austenitic stainless steel X5CrNi18-10. *Procedia CIRP* 3: 97–102.

30 Yadav, A.K., Bajpai, V., Singh, N.K., and Singh, R.K. (2017). FE modeling of burr size in high-speed micro milling of Ti6Al4V. *Precis. Eng.* 49: 287–292.

31 Saptaji, K., Subbiah, S., and Dhupia, J.S. (2012). Effect of side edge angle and effective rake angle on top burrs in micro milling. *Precis. Eng.* 36 (3): 444–450.

32 Kou, Z., Wan, Y., Cai, Y. et al. (2015). Burr controlling in micro milling with supporting material method. *Procedia Manuf.* 1: 501–511.

33 Kumar, P., Kumar, M., Bajpai, V., and Singh, N.K. (2017). Recent advances in characterization, modeling and control of burr formation in micro-milling. *Manuf. Lett.* 13: 1–5.

34 Chen, W. et al. Finite element simulation and experimental investigation on cutting mechanism in vibration assisted micro-milling. *Int. J. Adv. Manuf. Technol.* 105: 4539–4549.

8

Surface Topography Simulation Technology for Vibration-Assisted Machining

8.1 Introduction

Recently, great interest has been shown in the fabrication of functional surfaces with periodic micro- and nanostructures due to their application in new optics, automotive, aerospace, biomedical, and power generation devices [1, 2]; this is due to the excellent abilities, such as reducing adhesion friction, improving lubricity, and specific optical properties [3–5]. There are numerous machining methods for the production of microstructures, such as micro-electro-mechanical systems (MEMSs) technology, mechanical machining, and special processing [6, 7]. However, all of these methods have significant drawbacks, such as high cost, low efficiency, and the resulting serious environmental pollution [7, 8]. Diamond machining is the most widely used method for manufacturing precision parts, and is also capable of fabricating various surface structures, which have been illustrated in a recent review [9]. It is concluded that the range of machinable structures is limited and machining efficiency is dependent on the machined structures. The development of a novel machining method for microstructures has been a common research focus.

The generation of surface texture using vibration-assisted machining was recently proposed, which can generate a specific texture on the machined surface more conveniently. Sajjady et al. [10] and Brehl et al. [11] fabricated a microstructure by using elliptical vibration-assisted turning. Tong et al. [12] investigated the vibration-assisted electrical discharge machining (EDM) machining process of microstructures with noncircular cross section. Suzuki et al. [13, 14] also described the fabrication processes for several kinds of simple micro/nano-textured patterns on hardened steel and tungsten alloy using elliptical vibration-assisted turning, as shown in Figure 8.1a,b. Kim and Loh [15] directly machined micro-patterns on nickel alloy and mold steel by vibration-assisted turning, as shown in Figure 8.1c. From their primary experimental works, the proposed amplitude-control sculpturing method was verified to be feasible. Xu et al. [16] has reviewed recent advances in ultrasonic-assisted machining for the fabrication of micro/nano-textured surfaces and studied the micro/nanosurface generation with different vibration directions, a typical surface morphology was shown in as shown in Figure 8.1d. To summarize the existing research, elliptical vibration-assisted cutting attracts increasing attention, due to its excellent machining performances.

Vibration Assisted Machining: Theory, Modelling and Applications,
First Edition. Lu Zheng, Wanqun Chen, and Dehong Huo.

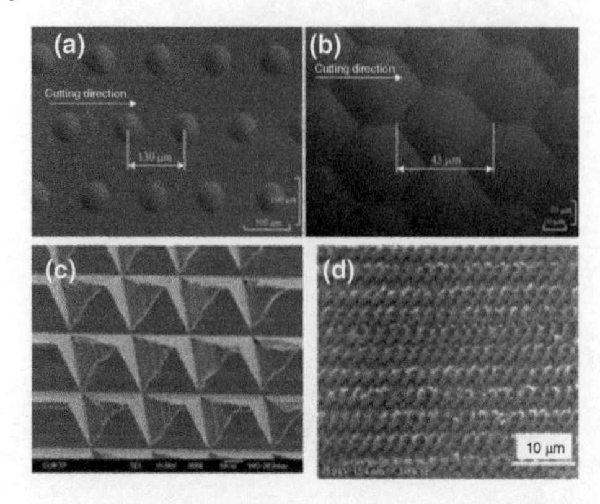

Figure 8.1 Typical structures fabricated with an elliptical ultrasonic texturing method. Source: (a, b) Suzuki et al. [13]. © 2011, Elsevier; (c) Kim and Loh [15]. © 2011, Springer Nature; (d) Xu et al. [16]. © 2017, Springer Nature.

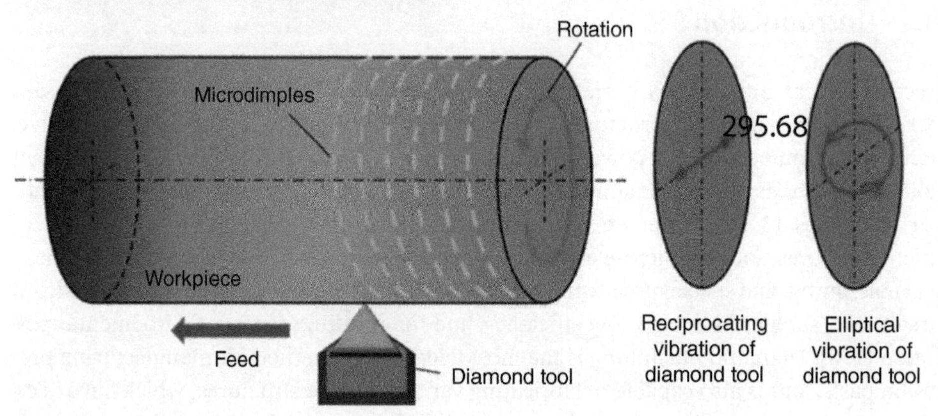

Figure 8.2 Ultrasonic-assisted turning processes for micro-texturing. Source: Xu et al. [16]. Licensed under CCBY 4.0.

Turning [12–14], grooving [17], and rotary machining [9–11] have been used to fabricate micro/nano-textured surfaces by integrating ultrasonic vibration motion. The cutting motion can be achieved with a rotating cylindrical workpiece, a linearly feed cutting tool, and a rotating cutting tool, respectively. The machinable structures achieved mainly depend on tool geometry, vibration mode, feed path, and their combinations. Figure 8.2 illustrates the principle of ultrasonic-assisted turning processes for micro-texturing [12, 13]. 1D and 2D vibration modes have been verified. When 1D vibration mode is applied, the continuous cutting process is transformed into an intermittent cutting process, allowing for the fabrication of microdimples. As for the 2D vibration mode, the authors developed a new elliptical ultrasonic vibration spindle that can make the tool vibrate in the cutting and depth-of-cut directions. By modulating the cutting depth and the vibration amplitude, a high-frequency intermittent contact between the cutting edge and workpiece is obtained, and this is possibly controlled for the fabrication of surface meso/micro-textures.

Besides, this vibration-assisted milling (VAMILL) has also recently been found to be capable of conveniently generating microstructures. Uhlmann et al. [18] found that surface topography and surface roughness in the process of micro-milling can be modulated by means of targeted high-frequency vibration of a workpiece. Tao et al. [19] realized the formation of surface texture by using ultrasonic VAMILL in the feed direction. They found that fish scale structure can be generated by VAMILL with optimal machining parameters. Börner et al. [20] found that by using defined cutting edge geometry and kinematics, the ultrasonic VAMILL represented a suitable method for the reproducible generation of a defined microstructure. Chen et al. [17] investigated a new surface texture formation method using nonresonant VAMILL, where two specific types of surface textures, wave and fish scale, are obtained by combining different machining parameters. VAMILL has exhibited great potential in the fabrication of microstructures, due to its high machining efficiency, low cost, and simple machine equipment setup. However, the surface texture generation process in vibration-assisted milling is very complicated; any change in processing parameters could result in significant change in the processed microstructure. Therefore, it is very important to investigate the influence of the machining and vibration parameters on the surface generation [17].

To solve this problem, such effort has been devoted to predicting the characteristic of surface generation in conventional machining and VAM. In order to investigate the influence of the machining process on the resulting surface, a description using analytical or geometric models is necessary. While analytical models represent mainly physical process properties by means of mathematical relationships, the geometrical models are used for the prediction of the geometric process properties. For this purpose, the modeling of the tool, the workpiece, and their relative movement is necessary. The main approach used for the simulation of a machined surface contains the stepwise intersection of the tool with the workpiece to subtract the overlapping volume from the workpiece model. Depending on the application, different modeling approaches exist for the representation of the intersection of the tool with the workpiece. Börner et al. [20] reviewed the surface modeling model in machining process, and Brecher et al. [21] distinguished between the volume-oriented constructive solid geometry (CSG) or boundary representations and the spatially discretized dexel or voxel representations, respectively. The CSG methods are frequently used for simulative purposes, as described from Weinert et al. [22]. The description of the machining process is achieved by Boolean intersection of CSG models, such as geometrical primitives in the form of cuboids, cylinders, and spheres. To reduce the computational effort, Klimant et al. [23] showed that the geometric material removal can also be represented by a sweep volume, in which the tool cross section is extruded along the tool path (trajectory) and then intersected with the tool. Both the workpiece and the tool can be presented in various ways. In general, however, the tools are simply represented as an envelope. Thus, an exact modeling of the resulting microstructure is only possible to a limited extent. This simplified surface representation is sufficient for many applications using the geometric information for the determination of cutting forces, temperatures, or collision calculation. However, for a precise modeling of the surface microstructure as a function of the tool geometry, a spatially partitioned prediction model, such as a dexel model, is more appropriate. In this model, a surface is represented by discrete points ("dexels") aligned over a regular grid in the X–Y plane and with a high-resolution Z-value. Nevertheless, the term dexel is used differently in literature. Altintas et al. [24] called such a set of unidirectional elements that share the same start value the z-buffer model.

Denkena and Böß [25] introduced an numerical control (NC) simulation system for research called CutS. It consists of a number of independent modules used to simulate the machine kinematics, material removal, the deformation of the workpiece, etc., and therefore it could offer information about the workpiece shape, contact zone, cutting forces, dynamic vibrations, and tool wear. The centerpiece of the software, the material removal module, used a CAD model for the tool and a three-dimensional dexel model (three perpendicular to each other arranged dexel models) for the workpiece that allowed the simulation of multiaxis cutting operations such as milling and grinding with high accuracy. Later Denkena et al. [26] used the simulation software for the simulation of the kinematic surface microstructure generated by ball end mills. Three-axis ball end milling is a popular field of investigation for surface simulation since the ball end contour is mathematically easier to depict compared to an end milling tool. Arizmendi et al. [27] and Piotrowska Kurczewski and Vehmeyer [28] presented two different analytical descriptions in the context of the modeling of the cutting edge geometry of ball end mills. For a more realistic display of the surface both took the tool run-out into account by defining a tool parallel axis offset. Denkena et al. [26] stated that the consideration of the cutting edge movement describes a more accurate method compared to a simple Boolean subtraction. Furthermore, he extended the common approach of simulating the kinematic surface microstructure by taking the stochastic topography into account. It characterizes the surface effects generated by tool deflection, current tool geometry, and stochastic influences. Therefore, a workpiece was experimentally machined with the same process parameters as the kinematic simulation. Afterwards the stochastic topography was generated by removing the computed topography from the measured workpiece surface using an empirical approach. The subsequent combination of both simulations created a real surface topography, which showed that high values of the feed per tooth and width of cut favor the dominance of the kinematic topography in relation to the stochastic topography. Another way to apply deterministic surface structures is high feed milling. This process is characterized by high feeds in combination with a low axial and radial depth. The special cutting geometry of the milling tool allows a higher feed, which leaves unmachined material sections on the finished surface. Consequently, this process results in a higher roughness. Freiburg et al. [29] simulated high feed milling to predict process forces, process dynamics, and surface structures. In their investigations, they compared a CSG modeled tool generated by scanning data from 3D light microscopy with a tool generated by an enveloping mesh. Although the first approach was more time consuming it delivered better simulation results in comparison to the experimental measured data: the simulation with the enveloping meshed tool model showed a difference in peak height but was still sufficient to qualitatively predict high feed surface structures much more rapidly.

For machining with an ultrasonic vibration assistance in the direction of the tool axis, such analytical investigations do not exist yet. However, surface prediction models are already available for other vibration-assisted machining methods. For example, Ding et al. [30] proposed a numerical modeling method for VAMILL process and a dexel-like calculation algorithm is proposed; in their model the tool profile was mapped along the tool path so that multiple tool engagements can be taken into account, but the effects of the tool geometry were overlooked in their investigation. Tao et al. [19] also analyzed the influence of tool parameters on the surface generation by simplifying the tool cutting edge as a sharp one, but this reduces the accuracy of the surface texture simulation. Börner et al. [20] proposed improved texture surface modeling for VAMILL, which is implemented by modeling both workpiece and tool in a dexel-based data model, and the actual tool geometry can be considered. On the whole, the current simulation method either has low simulation accuracy or has complicated calculation steps; a simple but accurate simulation method of VAMILL is desperately needed. In addition, as yet, the influences of the machining and vibration parameters on the textured surface generation are still not clear. It has become a significant technical obstacle preventing the extensive use of structured surface in creating innovative products when employing VAMILL.

In-depth understanding of the machining process of structured surface in vibration-assisted machining, especially knowing the relationship between machining and vibration parameters, surface texture, and its wettability, is an effective way to realize the deterministic manufacturing of functional surface. For these reasons, this chapter proposed an accurate modeling method to predict the surface topography in VAM process, which considers the geometrical features of cutting edge, along with the machining kinematics and coordinate transformation. The proposed method could realize the surface modeling with any tool geometry and complex machining trajectory. The feasibility of the proposed approach has been verified by milling experiments. The relationship between vibration parameters and surface texture generation is investigated. Furthermore, the effect of machined surface topography on its wettability performance is established. It clearly indicates that a controllable wettability surface can be generated using VAMILL if appropriate specific machining parameters are chosen.

This chapter is organized as follows. Section 8.2 gives the details of the modeling method for the generation of surface topography in VAMILL process. Section 8.2 introduces the VAMILL experiments to verify the modeling method proposed in Section 8.3. Moreover, in Section 8.4, the effects of the vibration parameters on surface wettability and the tool wear in the VAMILL process are analyzed and discussed. Finally, Section 8.5 summarizes the main conclusions of the present work.

8.2 Surface Generation Modeling in Vibration-Assisted Milling

The process of the generation of a textured surface can be recognized to be the result of the interactions between a cutter with a specific shape and workpiece. In the machining process, the cutter is fed along a specified machining path. Owing to the overlap of the trajectories of cutting edge on the workpiece surface, surface microstructures are formed on the machined surface. Thus, the surface generation in VAMILL is easily affected by various factors, including the kinematics of machining systems, the real geometry of the cutting edge, machining parameters, etc. To accurately model the surface generation in VAMILL, the detection and modeling of the cutter edge, the modeling and analysis of the kinematics, and the coordinate transformation from cutting edge to workpiece surface are the key steps.

As illustrated in Figure 8.3, a homogeneous matrix transformation (HMT)-based calculation algorithm for the surface generation of VAMILL is proposed, which mainly consists of the following steps:

Firstly, the geometry of the cutting edge is detected using scanning electron microscope (SEM) and atomic force microscope (AFM), and it can be described with a numerical function. Secondly, the cutting edge and workpiece are discretized into series of points. Thirdly,

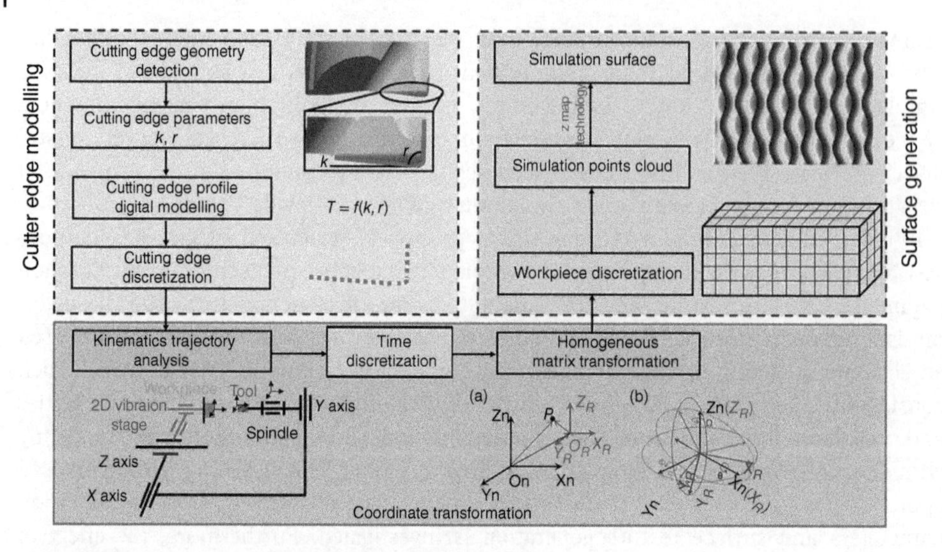

Figure 8.3 Flow chart of the surface simulation process.

Table 8.1 Parameters of the micro-cutting tool.

Parameter	Value
Cutting diameter (D)	0.5 mm
Number of flutes (N)	2
Cutter edge radius (r_e)	3 μm
Minor cutter edge angle (k)	5°
Corner radius (r)	5 μm

all points on the cutting edge are transferred from tool coordinate system to workpiece coordinate system by HMT, which forms the simulated surface in a point cloud. Then the Z-map technology is applied to plot the minimum value of z as a function of x and y in the workpiece coordinate system. Finally, the machined surface can be obtained.

8.2.1 Cutter Edge Modeling

To accurately predict the surface topography in the VAMILL process, the real contour of the cutter edge should first be described. Thus the micro-end mills were first detected by SEM and AFM before machining experiments in order to obtain the cutter geometry parameters. Figure 8.2a,b show the SEM images for the side and top views of the micro-milling tool. The minor cutter edge angle is measured with an analysis software ImageJ®. Besides, the cutter corner radius is detected by AFM, and the corner radius is further obtained by using the least squares fitting method, as shown in Figure 8.2c. The detailed geometrical parameters of the micro-cutting tool are summarized in Table 8.1.

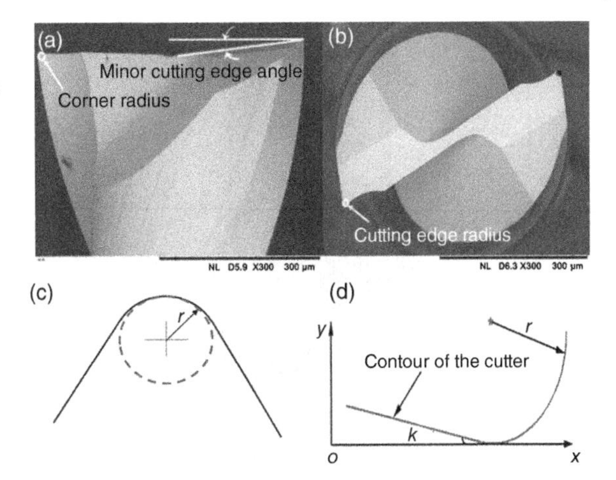

Figure 8.4 Tool geometry. (a) Side view; (b) top view; (c) corner radius fit; (d) end cutter edge profile. Source: Chen et al. [31]. © 2019, Elsevier.

The profile of the cutter edge, which is shown in Figure 8.4d, can be described as

$$z = f(x) = \begin{cases} -x \tan k + (R - r - r \sin k) \tan k + r(1 - \cos k), & 0 < x < R - r - r \sin k \\ -\sqrt{r^2 - [x - (R - r)]^2} + r, & x \geq R - r - r \sin k \end{cases} \tag{8.1}$$

where r, k, and R denote the tool corner radius, rake angle, and radius, respectively. The tool center is set to the origin of the coordinate system, Z_t axis is aligned with the tool axis, and the XY-plane is aligned with the tool rake face of one cutting flute.

8.2.2 Kinematics Analysis of Vibration-Assisted Milling

In VAMILL, in order to realize the oscillating motion between the tool and workpiece, high-frequency vibration with small amplitude could be superimposed on the tool or workpiece. In this research, vibration is applied to the workpiece. As shown in Figure 8.5, workpiece is simultaneously fed in the x- and y-direction, and the axial depth of cut is in the z-direction.

The relative displacement (x_i, y_i) of tool tip to workpiece in VAMILL can be depicted by Eq. (8.2):

$$\begin{cases} x_i = ft + r \sin\left[\omega t - \dfrac{2\pi(Z_i - 1)}{Z}\right] + A \sin(f_x t + \emptyset_x) \\ r \cos\left[\omega t - \dfrac{2\pi(Z_i - 1)}{Z}\right] + B \sin(f_y t + \emptyset_y) \end{cases} \tag{8.2}$$

where r and ω are the radius and angular velocity of the cutter, z_i is the ith cutter tooth, and Z is the number of flutes. f is the feed velocity. A and B are the vibration amplitudes, f_x and f_y are the vibration frequencies, and φ_x and φ_y are the phase angles, in the x- and y-directions.

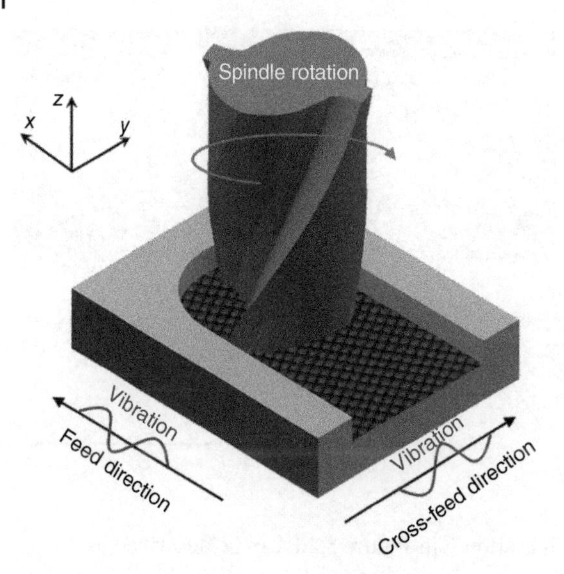

Figure 8.5 Schematic diagram of vibration-assisted milling.

8.2.3 Homogeneous Matrix Transformation

From Eq. (8.2), it can be noted that apart from t, 10 independent parameters are required to determine the tool tip trajectory in the 2D VAMILL. Owing to the effects of so many parameters, the tool tip trajectory in VAMILL is much more complex. In order to accurately describe the tool tip trajectory in VAMILL, the HMT method is adopted and the detailed HMT process in VAMILL is introduced in the following section.

8.2.3.1 Basic Theory of HMT

To represent the position of a rigid body in a three-dimensional space with a given coordinate system, a 4×4 matrix, ${}^{j}T_{i}$, is needed. This matrix represents the coordination transformation from the rigid body frame (o_i–x_i y_i z_i) to the reference coordinate frame (o_j–x_j y_j z_j), i.e. HMT. The first three columns of the homogeneous transformation matrix (HTM) refer to the direction cosines (unit vectors i, j, k). They represent the orientations of x_i, y_i, and z_i axes of the rigid body in the reference coordinate frame, respectively, all having scale factors that are zero. The last column represents the relative position of the coordinate system of the rigid body to the reference coordinate frame [17, 21].

Figure 8.4 shows the typical coordinate transformations from (o_i-x_i y_i z_i) to (o_j-x_j y_j z_j) including the translation transformation along x, y, and z, and the rotation transformation with α, β, and γ. The corresponding HTM is given in Table 8.2.

8.2.3.2 Establishment of HTM in the End Milling Process

To determine the effect of the error for each machine component on the relative position of tool tip to workpiece, the tool–workpiece spatial relationship should be predefined. To illustrate the relative position of a rigid body with respect to a given coordinate system, a 4×4 homogeneous transformation matrix (HTM, ${}^{R}T_{n}$) is needed. This also indicates the coordinate transformation from the rigid body frame (O_n-X_n Y_n Z_n) to the reference coordinate system (O_R-X_R Y_R Z_R).

Table 8.2 HTM from (*oi-xi yi zi*) to (*oj-xj yj zj*).

Translation transformation	***X*, *Y*, *Z***	${}^{j}T_{i}=\begin{bmatrix}1&0&0&x_{ij}\\0&1&0&y_{ij}\\0&0&1&z_{ij}\\0&0&0&1\end{bmatrix}$
Rotation transformation	α	${}^{j}T_{i}(\alpha)=\begin{bmatrix}1&0&0&0\\0&\cos\theta&-\sin\theta&0\\0&\sin\theta&\cos\theta&0\\0&0&0&1\end{bmatrix}$
	β	${}^{j}T_{i}(\beta)=\begin{bmatrix}\cos\varphi&0&-\sin\varphi&0\\0&1&0&0\\\sin\varphi&0&\cos\varphi&0\\0&0&0&1\end{bmatrix}$
	γ	${}^{j}T_{i}(\gamma)=\begin{bmatrix}\cos\Omega&-\sin\Omega&0&0\\\sin\Omega&\cos\Omega&0&0\\0&0&0&0\\0&0&0&1\end{bmatrix}$

In the HTM, the first three columns are direction cosines (unit vectors *i*, *j*, *k*). They represent the orientations of X_n, Y_n, and Z_n axes of the rigid body with respect to the (O_n-X_n Y_n Z_n), and their scale factors are zero. The last column in the HTM represents the relative position between the origins (O_n, O_R) of (O_n-X_n Y_n Z_n) and (O_R-X_R Y_R Z_R). Besides this, for clarity, P_s (a scale factor) is usually set to unit length.

As shown in Figure 8.6a, the transformation of the coordinate of point *P*(*x*, *y*, *z*) from (O_n-X_n Y_n Z_n) to (O_R-X_R Y_R Z_R) can be expressed as

$$\overrightarrow{O_nP}=\overrightarrow{O_nO_R}+\overrightarrow{O_RP}=(x,y,z)+(x_{Rn},y_{Rn},z_{Rn})=(x+x_{Rn},y+y_{Rn},z+z_{Rn}) \tag{8.3}$$

where (x_{Rn}, y_{Rn}, z_{Rn}) refers to the coordinates of O_n in (O_R-X_R Y_R Z_R).

The pre-superscript and post-subscript are applied to represent the reference frame and the reference frame after transformation, respectively. The transfer matrix is expressed as

$${}^{R}T_{n}=\begin{bmatrix}1&0&0&x_{Rn}\\0&1&0&y_{Rn}\\0&0&1&z_{Rn}\\0&0&0&1\end{bmatrix} \tag{8.4}$$

Thus, the coordinate transformation process of any point in a coordinate frame (n) to a reference frame (R) can be depicted via Eq. (8.3):

$$\begin{bmatrix} X_R \\ Y_R \\ Z_R \\ 1 \end{bmatrix} = {}^{R}T_n \begin{bmatrix} X_n \\ Y_n \\ Z_n \\ 1 \end{bmatrix} \tag{8.5}$$

In Figure 8.7b, it can be seen that $O_n(x, y, z)$ has been transformed to $O_R(x, y, z)$ after rotation around the x-, y-, and z-axis. The rotation angles have been labeled to be θ, φ, and Ω, respectively. To describe the transformation process from (O_n-X_n Y_n Z_n) to (O_R-X_R Y_R Z_R), which is realized by the rotation around x-axis with a rotation angle θ, the following equation is established.

$$\begin{cases} x_R = x_n \\ y_R = y_n \cos\theta - z_n \sin\theta \\ z_R = y_n \sin\theta + z_n \cos\theta \end{cases} \tag{8.6}$$

Thus, the corresponding homogeneous matrix could be expressed as

$$\begin{bmatrix} 1 & 0 & 0 & 0 \\ 0 & \cos\theta & -\sin\theta & 0 \\ 0 & \sin\theta & \cos\theta & 0 \\ 0 & 0 & 0 & 1 \end{bmatrix} \tag{8.7}$$

Similarly, the matrixes used to depict the rotation around y-axis with a rotation angle φ and the rotation around z-axis with a rotation angle Ω can be also expressed as

$$\begin{bmatrix} \cos\varphi & 0 & -\sin\varphi & 0 \\ 0 & 1 & 0 & 0 \\ \sin\varphi & 0 & \cos\varphi & 0 \\ 0 & 0 & 0 & 1 \end{bmatrix} \tag{8.8}$$

and

$$\begin{bmatrix} \cos\Omega & -\sin\Omega & 0 & 0 \\ \sin\Omega & \cos\Omega & 0 & 0 \\ 0 & 0 & 0 & 0 \\ 0 & 0 & 0 & 1 \end{bmatrix} \tag{8.9}$$

The HMT method has been explained in detail in Ref. [20] and it has been proved to be an effective tool for error analysis of machine tools [17, 21–23]. In the previous HMT model, it is recognized that the cutting trajectory of the machine tool can be generated by using one point, e.g. tool tip. In this research, the whole cutting edge profile is discretized into numerous points. Then, by combining with the Z-map technology, the 3D machined surface topography can be simulated.

8.2.3.3 HMT in VAMILL

It is widely recognized that the surface topography of a machined surface can be generated by the cutting of the cutter edge on the workpiece. To obtain the machined surface, the

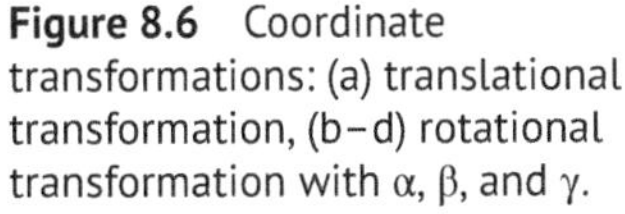

Figure 8.6 Coordinate transformations: (a) translational transformation, (b–d) rotational transformation with α, β, and γ.

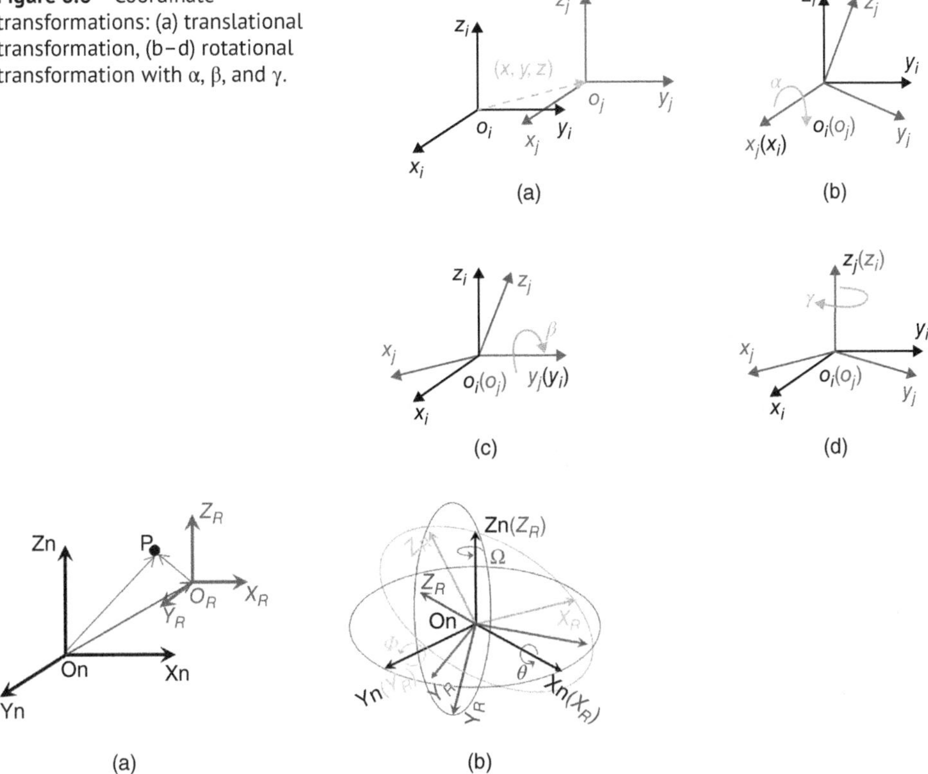

Figure 8.7 Coordinate transformations: (a) translational transformation; (b) rotational transformation.

HMT of the VAMILL should be conducted for the whole cutter edge. The HMT of a single point transformation from (o_i-x_i y_i z_i) to (o_j-x_j y_j z_j), is illustrated in Figure 8.7, and the tool tip trajectory can be obtained using this method.

Figure 8.8 shows a schematic diagram of the VAMILL equipment setup. The removal of workpiece material is performed by the cutter with high rotation speed. The vibration is applied to the workpiece, which is fed along the x-direction. Also, it can be seen that the local coordinate system of the cutter (O_t-$X_tY_tZ_t$) and the rotation coordinate system (O_T-$X_TY_TZ_T$) are applied to describe the tool profile and the spindle rotation. Besides this, the translation coordinate systems, i.e. O_V-$X_VY_VZ_V$ and O_W-$X_WY_WZ_W$, are used to depict the vibration on the workpiece and the feed of the control system. The detailed transformation or rotation values of each coordinate system in the VAMILL process are given in Table 8.3.

The homogeneous coordinates of the cutter edge in O_t-$X_tY_tZ_t$ are given as

$$\begin{bmatrix} x_t \\ y_t \\ z_t \\ 1 \end{bmatrix} = {}^{R}T_n \begin{bmatrix} x \\ 0 \\ f(x) \\ 1 \end{bmatrix} \tag{8.10}$$

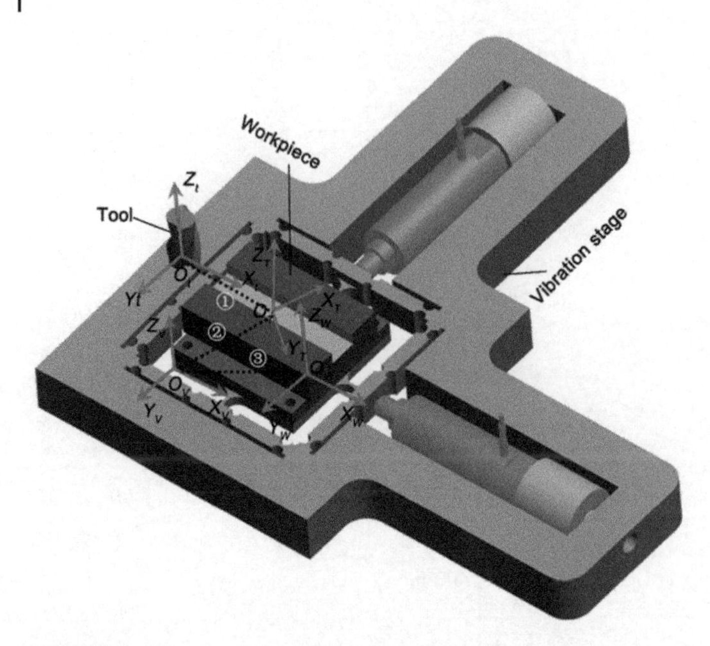

Figure 8.8 The coordinate systems in vibration-assisted milling.

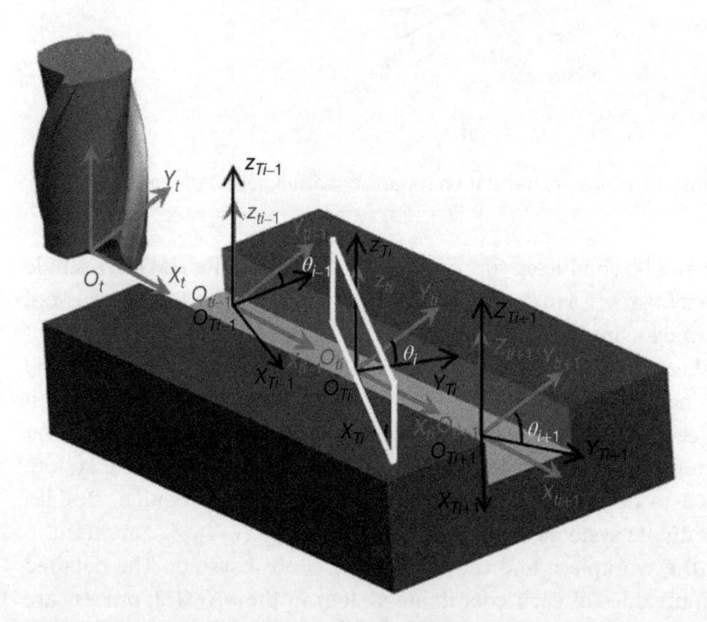

Figure 8.9 Tool instantaneous attitude position.

The O_T-$X_TY_TZ_T$ only considers the rotation speed of the cutter. The X_T and Y_T axes are aligned with the cutter feed and cross-feed directions, respectively. The Z_T axis is parallel to Z_t. The origin of the tool coordinate system is the same as that of the local tool coordinate system. The angular velocity of O_t-$X_tY_tZ_t$ relative to O_T-$X_TY_TZ_T$ is therefore the spindle angular velocity. As shown in Figure 8.6, the relative rotational angle (θ) of two coordinate

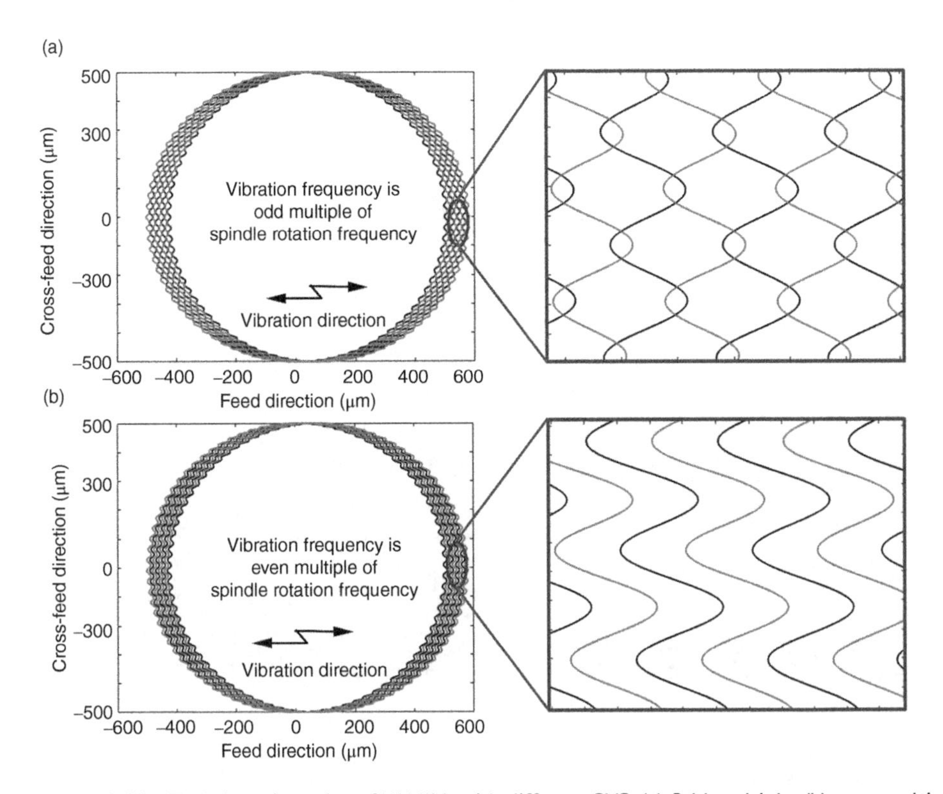

Figure 8.10 Tool tip trajectories of VAMILL with different RVS. (a) Odd multiple, (b) even multiple.

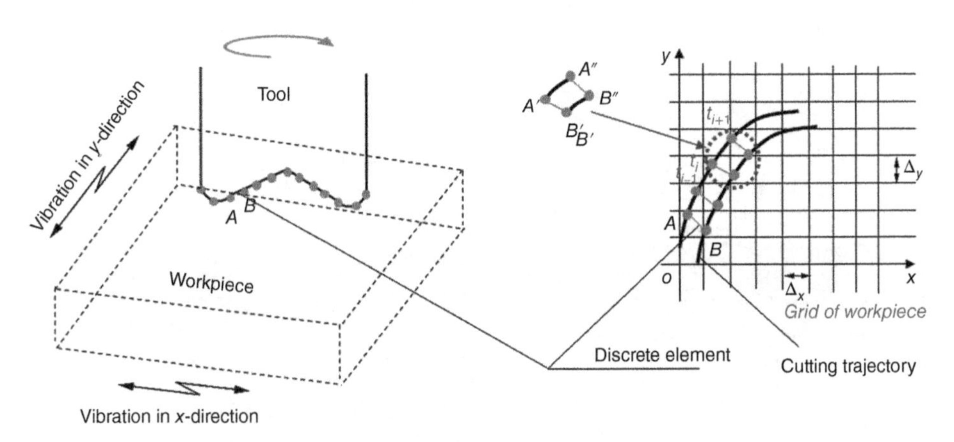

Figure 8.11 Schematic of surface generation simulation algorithm.

systems changes with time. When the points of the cutter edge are changed from O_t-$X_tY_tZ_t$ to O_T-$X_TY_TZ_T$, the rotational effect is considered.

$$\theta = \phi - \omega t \tag{8.11}$$

where ϕ is the initial angle between X_t and X_T axes. In this research, it is set to 0°. ω is the tool rotation angular velocity (rad/s) and t is the cutting time (s).

Figure 8.12 Simulation results of the VAM surface. (a–h) Surface generated with the machining and vibration parameters of Set 1–8.

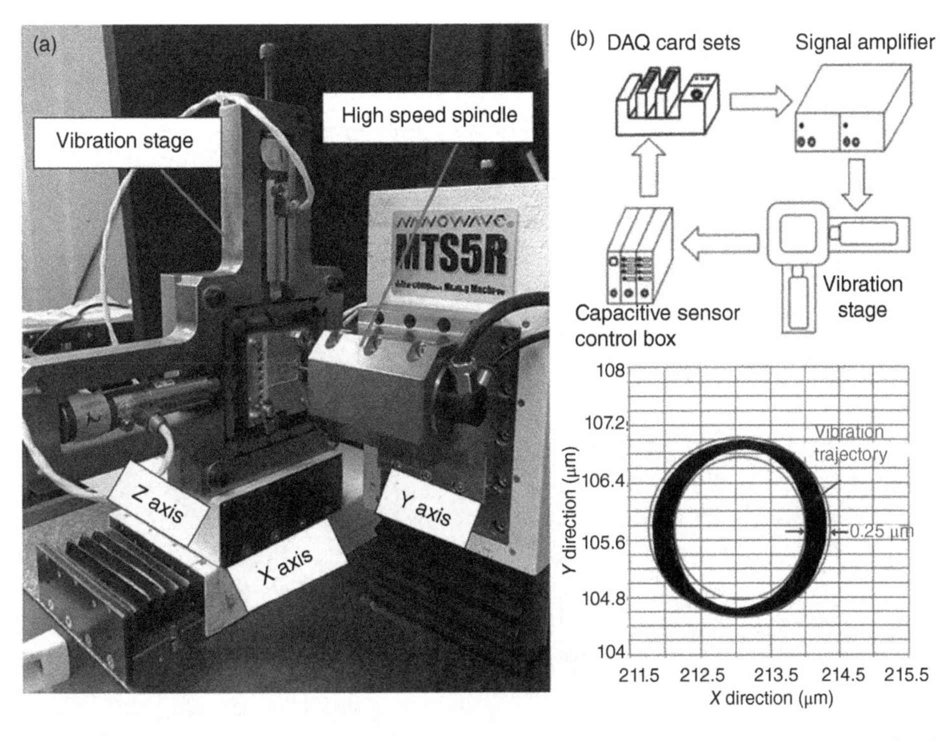

Figure 8.13 Layout of VAMILL equipment. (a) Vibration-assisted milling system; (b) schematic of the vibration stage control system; (c) error test result of the vibration stage. Source: Chen et al. [31]. © 2019, Elsevier.

Table 8.3 State value for each HMT.

HMT	x	y	z	α	β	γ
$^{T}T_{t}$	0	0	0	0	0	$\phi - \omega t$
$^{V}T_{T}$	$A\sin(2\pi f_x t + \emptyset_x)$	$B\sin(2\pi f_y t + \emptyset_y)$	0	0	0	0
$^{W}T_{V}$	$x_0 + (i-1)a_e$	$y_0 + v_f t$	0	0	0	0

The HTM from O_t-$X_tY_tZ_t$ to O_T-$X_TY_TZ_T$ can be expressed as

$$^{T}T_{t} = \begin{bmatrix} \cos\theta & -\sin\theta & 0 & 0 \\ \sin\theta & \cos\theta & 0 & 0 \\ 0 & 0 & 1 & 0 \\ 0 & 0 & 0 & 1 \end{bmatrix} \tag{8.12}$$

In the workpiece coordinate system (O_W-$X_WY_WZ_W$), X_W and Y_W refer to the feed direction and cross-feed direction, respectively. Z_W represents the direction of the axial depth of cut, which is parallel to Z_t. In a three-axis milling process, only the translation between O_W-$X_WY_WZ_W$ and O_T-$X_TY_TZ_T$ occurs. The initial position of O_W-$X_WY_WZ_W$ relative to O_T-$X_TY_TZ_T$ could be determined according to the actual machining situation.

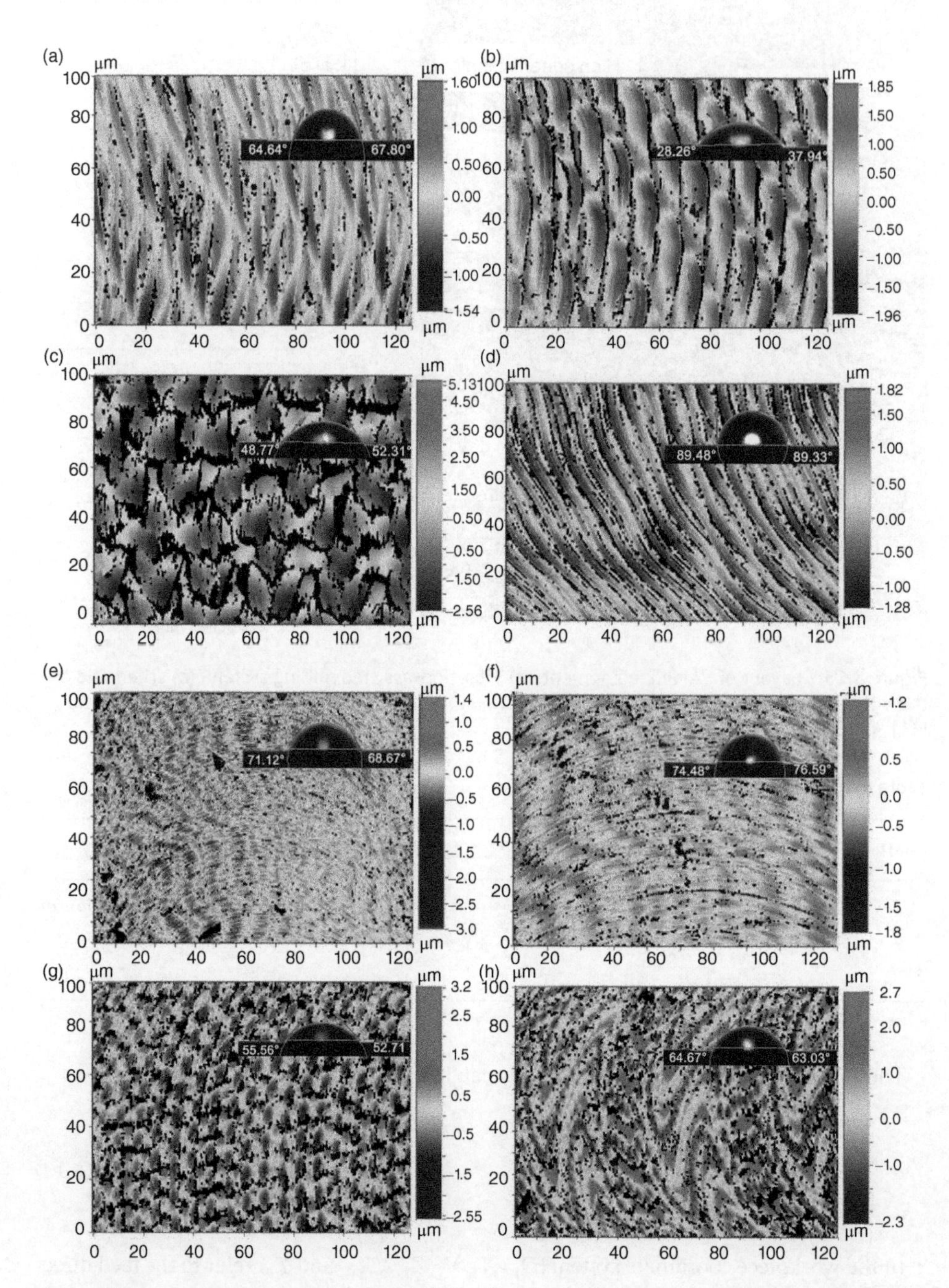

Figure 8.14 Experimental results of the VAM surface. (a–h) Surface generated with the machining and vibration parameters of Set 1–8.

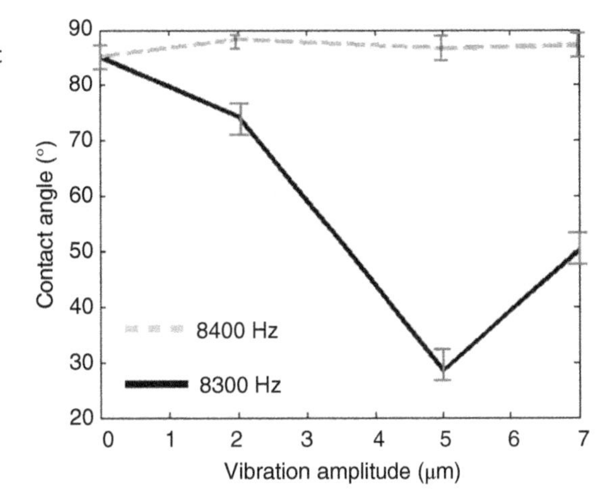

Figure 8.15 Influence of the vibration frequency and amplitude on the contact angle.

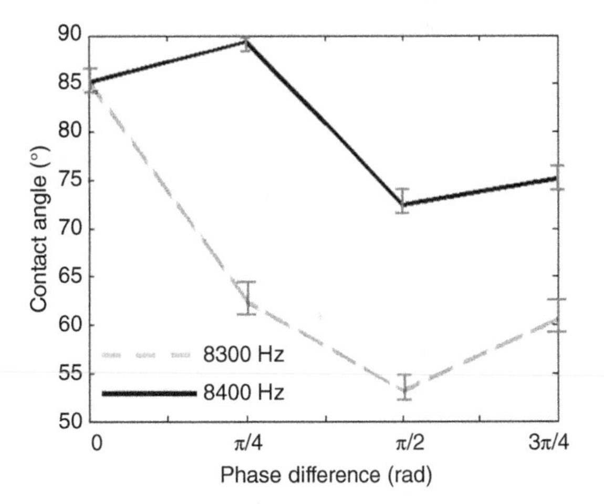

Figure 8.16 Influence of the phase difference and vibration frequency on the contact angle.

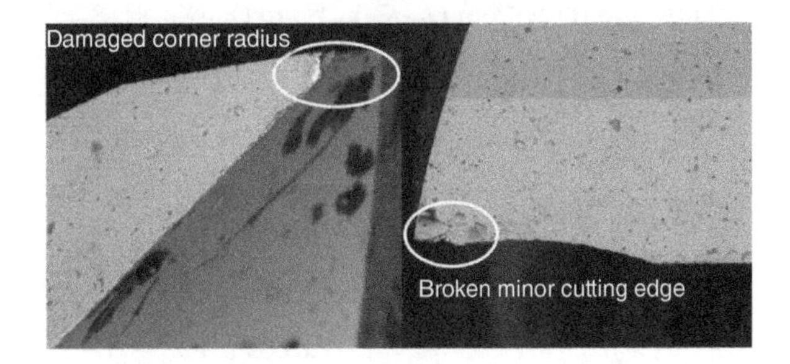

Figure 8.17 Tool wear for fish scale generation. Source: Chen et al. [31]. © 2019, Elsevier.

By transforming the coordinates of the tool cutter edge from O_t-$X_tY_tZ_t$ to O_W-$X_WY_WZ_W$, the position of each point on the cutter edge relative to the workpiece can be obtained, and the height of the cutter edge at any point could also be recorded through the Z_W coordinate.

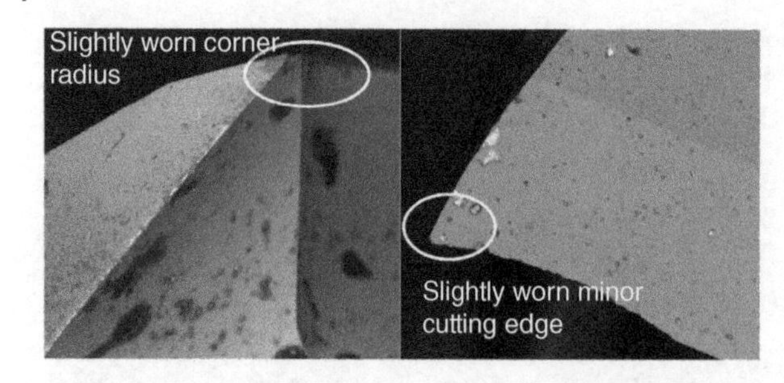

Figure 8.18 Tool wear for wave surface generation. Source: Chen et al. [31]. © 2019, Elsevier.

The original coordinate of O_T-$X_TY_TZ_T$ for each feed in cross-feed direction relative to O_W-$X_WY_WZ_W$ is obtained by

$$\begin{cases} x_{oT} = x_o + (i-1)a_e \\ y_{oT} = y_o + v_f t \\ z_{oT} = z_o \end{cases} \tag{8.13}$$

where (x_0, y_0, z_0) is the initial position of the origin of O_T-$X_TY_TZ_T$ in the O_W-$X_WY_WZ_W$, which is set as $(0, -R, 0)$ in this research. a_e is the radial cutting depth (mm). i is the number of feed in cross-feed direction; v_f is the feed speed in Y_w direction (mm/s).

As shown in Figure 8.9, owing to the effect of the vibration applied on the workpiece, the relative displacement between tool and workpiece is changed. Thus, the origin coordinates of O_T-$X_TY_TZ_T$ for each feed in radial direction relative to the origin of O_W-$X_WY_WZ_W$ can be calculated as

$$\begin{cases} x_{oT} = x_o + (i-1)a_e + A\sin(f_x t + \emptyset_x) \\ y_{oT} = y_o + v_f t + B\sin(f_y t + \emptyset_y) \\ z_{oT} = z_o \end{cases} \tag{8.14}$$

where $x(t)$ and $y(t)$ denote the components of the dynamic instantaneous displacement between tool and workpiece in the x- and y-direction, respectively.

Thus, the HTM from O_T-$X_TY_TZ_T$ to O_W-$X_WY_WZ_W$ can be expressed as

$${}^{W}T_T = \begin{bmatrix} 1 & 0 & 0 & x_o + (i-1)a_e + A\sin(f_x t + \emptyset_x) \\ 0 & 1 & 0 & y_o + v_f t + B\sin(f_y t + \emptyset_y) \\ 0 & 0 & 1 & z_o \\ 0 & 0 & 0 & 1 \end{bmatrix} \tag{8.15}$$

The cutter edge in the workpiece coordinate system O_W-$X_WY_WZ_W$ can be expressed as

$$\begin{bmatrix} x_W \\ y_w \\ z_w \\ 1 \end{bmatrix} = {}^{W}T_T{}^{T}T_t \begin{bmatrix} x_t \\ y_t \\ z_t \\ 1 \end{bmatrix} = \begin{bmatrix} x\cos\theta + x_o + (i-1)a_e + A\sin(f_x t + \emptyset_x) \\ x\sin\theta + y_o + v_f t + B\sin(f_y t + \emptyset_y) \\ f(x) + z_o \\ 1 \end{bmatrix} \tag{8.16}$$

Figure 8.10 illustrates two typical tool tip trajectories in VAMILL based on the HMT. When the vibration is an odd multiple of the spindle rotation frequency, a complex surface texture is generated by the overlapping between the adjacent cuter teeth. When the vibration is an even multiple of spindle rotation frequency, a wave surface is generated. It can be noted that the ratio between the vibration frequency and the spindle rotation speed (RVS) has a significant influence on the texture generation.

8.2.4 Surface Generation

According to the copy principle of tool contours, the workpiece surface topography is formed by sweeping the cutter edge profile along the tool path.

In order to model the surface generation through numerical simulations, cutter edge and workpiece are discretized into a series of elements. The machined surface topography could be formed by mapping the cutter edge contour onto the workpiece surface. Figure 8.11 illustrates the schematic diagram of the simulation of surface generation. Firstly, the cutter edge is discretized into numerous elements, and the workpiece is evenly divided into n columns and m rows. The intersection of each row and column forms a grid point. The height of each point on the machined workpiece surface can be obtained by calculating the tool cutting depth (z-coordinate) at grid point. To obtain one complete surface, the height of the middle position between two grid points can be obtained using a curve-fitting algorithm.

8.2.4.1 Surface Generation Simulation

In-depth understanding of the machining process of structured surface in VAMILL, especially when the relationship between machining and vibration parameters, the surface texture, and its wettability are known, makes surface simulation an effective way to realize the deterministic manufacturing of functional surfaces.

In order to investigate the influence of vibration parameters on surface generation, the machining and cutter parameters, such as spindle rotation speed, number of teeth, and the feed per teeth, are predefined as shown in Table 8.4. A previous investigation [18] shows that the ratio of vibration frequency to spindle rotation speed, as well as vibration amplitude significantly influence on the texture generation. To clearly indicate the influence of vibration parameters on the surface topography, eight typical sets of vibration parameters are selected in this study. The spindle speed is kept at 6000 rpm (rotation frequency is 100 Hz). The vibration frequency is set to 8300 Hz (odd multiple of spindle rotation frequency) and 8400 Hz (even multiple of spindle rotation frequency). Meanwhile, three typical vibration amplitudes, that is, 2, 5, and 7 μm, are employed. They are less, equal, and larger than half of the feed per tooth (10 μm), respectively. In addition, to investigate the influence of a phase difference between the two vibration directions on the surface generation, the phase difference is set to 0, $\pi/4$, and $\pi/2$, respectively.

From Set 1–3 (see Table 8.4), it can be found that the frequency of the vibration applied in x-direction is an odd multiple of the spindle rotation frequency. Since the applied vibration trajectory is a sinusoidal curve, the valleys (crests) of the current cutter trajectory are intersected with the crests (valleys) of the following cutter trajectory. As a result, three types of surface topographies are obtained, as shown in Figure 8.12a–c. It can be found that different

Table 8.4 Machining and vibration-assisted parameters.

			x-direction		y-direction		
Set no.	**Spindle speed (rpm)**	**Feed per tooth (μm)**	**Vibration amplitude (μm)**	**Vibration frequency (Hz)**	**Vibration amplitude (μm)**	**Vibration frequency (Hz)**	**Phase difference (rad)**
1	6000	10	2	8300	0	0	0
2	6000	10	5	8300	0	0	0
3	6000	10	7	8300	0	0	0
4	6000	10	7	8400	0	0	0
5	6000	10	5	8400	5	8400	$\pi/2$
6	6000	10	5	8400	5	8400	0
7	6000	10	5	8300	5	8300	0
8	6000	10	5	8300	5	8300	$\pi/4$

kinds of fish squamous structures are generated on the machined surface, corresponding to the vibration amplitudes of less, equal, and larger than half of the feed per tooth.

In the VAMILL experiments of Set 4, since the vibration frequency applied in the x-direction is an even multiple of the spindle rotation frequency, the crests (valleys) of the current cutter trajectory are intersected with those of the following cutter trajectory, thereby leading to the formation of a wavy surface, as shown in Figure 8.12d.

In the milling experiments of Sets 5 and 6, the vibration is applied to both the x- and y-directions. Moreover, the vibration frequencies are an even multiple of the spindle rotation frequency, and the phase differences between two vibration signals are $\pi/2$ and 0, respectively. Under the effects of the x- and y-direction vibrations, it can be seen in Figure 8.12e,f that the crests (valleys) of the current cutter trajectory also meet the crests (valleys) of the following cutter trajectory. Meanwhile, the spatial distribution of wavy surface is changed, due to the phase difference between two vibration signals, as shown in Figure 8.12e,f.

Finally, similar to those of Sets 5 and 6, vibrations are also applied in x- and y-directions in the milling experiments of Sets 7 and 8. However, the vibration frequencies (odd multiple of spindle rotation frequency) and the phase differences of zero and $\pi/2$ between two vibration signals are adopted. From the results (see Figure 8.12g,h), it can be noted that the valleys (crests) of the current cutter trajectory also meet the crests (valleys) of the following cutter trajectory. Finally, two evidently different surface topographies are finally obtained, which can be attributed to the effects of different phase differences between the two vibration signals.

8.3 Vibration-Assisted Milling Experiments

In the present work, all of the milling experiments were carried out on a three-axis precision milling machine tool (NANOWAVE MTS5R) equipped with a 2D vibration stage on the z-direction machine guideway, as shown in Figure 8.13a. Figure 8.13b shows a schematic of the vibration stage control system. The control signals are set by a host computer and amplified through a high-voltage piezo amplifier. They are then used to drive the piezo actuators. Meanwhile, the stage displacement data is detected by two high-precision capacitive sensors (CS005, Micro-epsilon) and fed back to the host computer through data acquisition cards for recording. Figure 8.13c shows an example of the test results, where two voltage signals (0.2 V, 1000 Hz) with a 90° phase difference are sent into the system. In the VAM control process, vibration trajectory errors are unavoidable due to stage manufacturing errors, interference from electronic devices, and control signal distortions. For the vibration stage used in this experiment, a total vibration trajectory error of 0.25 μm was measured. The uncoated double-edged micro-tool is adopted, and its diameter is 0.5 mm. The workpiece material is Al-6061. In order to verify the above simulation results, the machining parameters used in the experiments are the same as those adopted in the simulation.

The surface topography of each machined surface was tested using a white light interferometer surface profilometer (Vecco NT1100), as shown in Figure 8.12. It can be clearly seen that the results seen in Figure 8.12a,b,d agree well with the simulation results (see Figure 8.9a–c), respectively. However, other experimental results are not consistent with the simulation results in Figure 8.9. This could be ascribed to the effects of different kinds of vibrations. In the milling experiments corresponding to Figure 8.12a,b,d, the vibration is only applied in one direction. The control of vibration amplitude is easy. However, in the milling experiments corresponding to Figure 8.12e–h, vibration is applied in both x- and y-directions. Owing to the stiffness coupling effect of 2D-vibration stage, due to the interaction of the vibrations in the x- and y-directions, vibration amplitude and the phase difference between x- and y-direction vibration signals do not reach the expected value. As for the milling corresponding to Figure 8.12c, due to the elastic recovery effect of the material, the overlap of two adjacent cutter edges is too small to be reflected on the machined surface.

Overall, the surface texture model proposed in the present work can accurately describe the surface generation in 1D VAMILL process. However, due to the coupling effect of the x- and y-direction vibration, it is still hard to accurately predict the surface generation in the 2D VAMILL process.

8.4 Discussion and Analysis

The water contact angle on each machined surface was measured by a Sindatek Water Contact Goniometer with 5 ml water droplets. From Figure 8.14, it can be also clearly seen that the workpiece surface generated with different vibration parameters exhibit different wettability. According to the measured water contact angles, the wettability of the fish scale surface is evidently better than that of the wavy structure. The practical contact angle can be controlled from 30° to 80° by changing the vibration parameters. Therefore, with VAMILL it is absolutely feasible to obtain the machined surface with controllable wettability, which could be realized by only adjusting the vibration parameters in the milling process.

8.4.1 The Influence of the Vibration Parameters on the Surface Wettability

To quantitatively analyze the influence of vibration frequency, amplitude, and phase difference on the surface wettability of the machined surface, Figure 8.15 plots the contact angle versus vibration frequency and amplitude when vibration is applied in the feed direction.

It can be noted that when the vibration frequency is 8400 Hz, the wave texture is generated on the surface and the texture has little effect on the contact angle, which means that the wave texture barely changes the hydrophobicity of the workpiece surface. Meanwhile, when the vibration frequency changes to 8300 Hz, which is an odd multiple of the spindle rotation frequency, the fish squamous structure is generated on the surface, which has a significant effect in reducing the contact angle. This means that the fish squamous structure can increase hydrophilicity and the contact angle decreases with increasing vibration amplitude, reaching the lowest point when the vibration amplitude is equal to half of the feed per tooth, and after that the contact angle shows an upward trend with increases in vibration amplitude.

Figure 8.15 illustrates the influence of vibration frequency and amplitude on the contact angle when the vibration is applied in the feed direction. It can be noted that when the vibration frequency is 8400 Hz, which is an even multiple of the spindle rotation frequency of 100 Hz, the wave texture is generated on the surface, and this texture has little effect on the contact angle; therefore, the wave texture hardly changes the hydrophobicity of the workpiece surface. However, when the vibration frequency changes to 8300 Hz, which is an odd multiple of the spindle rotation frequency, the fish squamous structure is generated on the surface, which has a significant effect in reducing the contact angle. This texture can increase hydrophilicity and the contact angle decreases with increase in the vibration amplitude, reaching the lowest point when the vibration amplitude is equal to half of the feed per tooth, after which the contact angle shows an upward trend with increasing vibration amplitude.

Figure 8.16 illustrates the influence of phase difference and vibration frequency on contact angle when the vibration is applied in the feed and cross-feed direction. It can be noted that when the vibration frequency is 8400 Hz (even multiples of the spindle rotation frequency), the contact angle shows a slight upward trend with a $\pi/4$ phase difference, and after that the contact angle is reduced compared with no vibration, and reaches the lowest point with a $\pi/2$ phase difference, and after that the contact angle increases but is still less than for the surface generated without vibration. Meanwhile, when the vibration frequency changes to 8300 Hz (odd multiples of the spindle rotation frequency), the complex surface textures are generated on the machined surface, which has the effect of reducing the contact angle. The contact angle decreases with increasing phase difference, and reaches the lowest point when the phase difference is equal to $\pi/2$; after that the contact angle shows an upward trend with increasing vibration amplitude, but still less than the surface generated without vibration.

8.4.2 Tool Wear Analysis

Tool wear was compared when the vibration is applied in the feed direction with different vibration frequencies. The two vibration frequencies are 8400 and 8300 Hz, which are even and odd multiples of the spindle rotation frequency of 100 Hz, respectively. The tools are detected after machining 30 mm × 20 mm slots, and it can be noted from Figure 8.14, the corner radius and minor cutting edge are damaged; the reason for this is that when the vibration frequency is an odd multiple of the spindle rotation frequency, the valleys (crests) of the current cutter trajectory meet the crests (valleys) of the following cutter trajectory, and the instantaneous cutting thickness changes drastically, causing an intermittent impact on the tool. In contrast, when the vibration frequency is an even multiple of the spindle rotation frequency, the valleys (crests) of the current cutter trajectory meet the valleys (crests) of the following cutter trajectory, and there is no sudden change in instantaneous cutting thickness and the cutting force changes smoothly. Slight wear was found in the corner radius and minor cutting edge, as shown in Figure 8.15. It can be noted that the ratio of vibration frequency to spindle rotation frequency has a significant influence on the surface texture generation, as well as on tool wear, for the generation of the fish scale structure; the tool wear occurs sooner than the wave structure (Figures 8.17 and 8.18).

8.5 Concluding Remarks

In this chapter, based on the HMT and cutter edge sweeping technology, a method for the modeling of surface topography in vibration-assisted micro-milling is proposed. The relationship between machining and vibration parameters and the contact angle is established, and the controllable wettability of the machined surface can be achieved by changing the vibration parameters.

References

1 Hadad, M. and Ramezani, M. (2016). Modeling and analysis of a novel approach in machining and structuring of flat surfaces using face milling process. *Int. J. Mach. Tools Manuf.* 105: 32–44.
2 Wilkinson, P., Reuben, R.L., Jones, J.D.C. et al. (1997). Surface finish parameters as diagnostics of tool wear in face milling. *Wear* 205: 47–54.
3 Jung, Y.C. and Bhushan, B. (2006). Contact angle, adhesion and friction properties of micro-and nanopatterned polymers for superhydrophobicity. *Nanotechnology* 17 (19): 4970.
4 Maboudian, R. and Howe, R.T. (1997). Critical review: adhesion in surface micromechanical structures. *J. Vac. Sci. Technol. B: Microelectron. Nanometer Struct. Process. Meas. Phenom.* 15 (1): 1–20.

5 Liao, H., Normand, B., and Coddet, C. (2000). Influence of coating microstructure on the abrasive wear resistance of WC/Co cermet coatings. *Surf. Coat. Technol.* 124 (2): 235–242.
6 Judy, J.W. (2001). Microelectromechanical systems (MEMS): fabrication, design and applications. *Smart Mater. Struct.* 10 (6): 1115.
7 Dornfeld, D., Min, S., and Takeuchi, Y. (2006). Recent advances in mechanical micromachining. *CIRP Ann. – Manuf. Technol.* 55 (2): 745–768.
8 Cheng, K. and Huo, D. (2013). *Micro-Cutting: Fundamentals and Applications.* Wiley.
9 Brinksmeier, E., Gläbe, R., and Schönemann, L. (2012). Review on diamond machining processes for the generation of functional surface structures. *CIRP J. Manuf. Sci. Technol.* 5 (1): 1–7.
10 Sajjady, S.A., Abadi, H.N.H., Amini, S. et al. (2016). Analytical and experimental study of topography of surface texture in ultrasonic vibration assisted turning. *Mater. Des.* 93: 311–323.
11 Brehl, D.E., Dow, T.A., Garrard, K. et al. (1999). Micro-structure fabrication using elliptical vibration-assisted machining (EVAM). *ASPE Proc.* 39: 511–515.
12 Hao, T., Li, Y., and Yang, W. (2008). Experimental research on vibration assisted EDM of micro-structures with non-circular cross-section. *J. Mater. Process. Technol.* 208 (1): 289–298.
13 Suzuki, N., Yokoi, H., and Shamoto, E. (2011). Micro/nano sculpturing of hardened steel by controlling vibration amplitude in elliptical vibration cutting. *Precis. Eng.* 35: 44–50.
14 Suzuki, N. et al. (2007). Elliptical vibration cutting of tungsten alloy molds for optical glass parts. *CIRP Ann. – Manuf. Technol.* 56 (1): 127–130.
15 Kim, G.D. and Loh, B.G. (2011). Direct machining of micro patterns on nickel alloy and mold steel by vibration assisted cutting. *Int. J. Precis. Eng. Manuf.* 12 (4): 583–588.
16 Xu, S. et al. (2017). Recent advances in ultrasonic-assisted machining for the fabrication of micro/nano-textured surfaces. *Front. Mech. Eng.* 12 (1): 33–45.
17 Chen, W. et al. (2018). Surface texture formation by non-resonant vibration assisted micro milling. *J. Micromech. Microeng.* 28 (2): 025006.
18 Uhlmann, E., Perfilov, I., and Oberschmidt, D. (2015). Two-axis vibration system for targeted influencing of micro-milling. In: *Euspen's 15th International Conference & Exhibition*, 325–326. Leuven, Belgium.
19 Tao, G. et al. (2017). Feed-direction ultrasonic vibration-assisted milling surface texture formation. *Mater. Manuf. Processes* 32 (2): 193–198.
20 Börner, R. et al. (2018). Generation of functional surfaces by using a simulation tool for surface prediction and micro structuring of cold-working steel with ultrasonic vibration assisted face milling. *J. Mater. Process. Technol.* 255: 749–759.
21 Brecher, C., Daniels, M., Wellmann, F. et al. (2015). *Realisierung Effizienter Zerspanprozesse (Ergebnisbericht Des BMBF Verbundprojekts ReffiZ).* Aachen: Shaker Verlag GmbH.
22 Weinert, K., Kersting, P., Surmann, T., and Biermann, D. (2008). Modeling regenerative workpiece vibrations in five-axis milling. *Prod. Eng. Res. Dev.* 2: 255–260.
23 Klimant, P., Witt, M., and Kuhl, M. (2014). CAD kernel based simulation of milling processes. *Procedia CIRP* 17: 710–715.

24 Altintas, Y., Kersting, P., Biermann, D. et al. (2014). Virtual process systems for part machining operations. *CIRP Ann.* 63: 585–605.

25 Denkena, B. and Böß, V. (2009). Technological NC simulation for grinding and cutting processes using CutS Spain, In: *Proceedings of the 12th CIRP Conference on Modelling of Machining Operations*, 563–566. Donostia-San Sebastián.

26 Denkena, B., Böß, V., Nespor, D., and Samp, A. (2011). Kinematic and stochastic surface topography of machined TiAl6V4-parts by means of ball nose end milling. *Procedia Eng.* 19: 81–87.

27 Arizmendi, M., Fernández, J., Lacalle, L.N.L.d. et al. (2008). Model development for the prediction of surface topography generated by ball-end mills taking into account the tool parallel axis offset. Experimental validation. *CIRP Ann. – Manuf. Technol.* 57: 101–104.

28 Piotrowska Kurczewski, I. and Vehmeyer, J. (2011). Simulation model for micro-milling operations and surface generation. *Adv. Mater. Res.* 223: 849–858.

29 Freiburg, D., Kersting, P., and Biermann, D. (2015). Simulation and structuring of complex surface areas using high feed milling. In: *8th International Conference and Exhibition on Design and Production of Machines and Dies/Molds*, 93–98. Kusadasi, Aydin, Türkei.

30 Ding, H., Chen, S.-J., and Cheng, K. (2010a). Dynamic surface generation modeling of two-dimensional vibration-assisted micro-end-milling. *Int. J. Adv. Manuf. Technol.* 53: 1075–1079.

31 Chen, W., Zheng, L., Xie, W. et al. Modelling and experimental investigation on textured surface generation in vibration-assisted micro-milling. *J. Mater. Process. Technol.* 266: 339–350.

Index

a

abnormal wear 95–96
  chipping 95–96
  cracking 95–96
  fracture 95–96
  spalling 95–96
abrasive wear 97
actuators 18–19
  magnetostrictive actuators 18
  piezoceramic rings 21
  piezoelectric actuators 18
adhesion wear 97
analytical rigid body 148
ANSYS 63
Arbitrary Lagrangian Eulerian (ALE) formulation 148
atomic force microscope (AFM) 171

b

block type chips 141
boundary conditions 54
brittle damage 95
burr 8
  burr reduction 8, 108–115

c

Cartesian coordinate system 60
Castigliano's second theorem 56
ceramic stack 46
chip generation 108
chip geometric feature 120
chip thickness 73, 89
  uncut chip thickness 106–107
  undeformed chip thickness 7, 112, 145
clamping mode 20, 35
coated tool 97
coating delamination 97
cold welding 97
compliance transformation matrix 60–63
composite horns 53
compound motion 34–35
compression shear deformation 112
constructive solid geometry 169
contact angles 188
continuous type chips 141
controllable wettability surface 171
coupling effect 31–33, 36–37, 45
crater 96
critical cutting depth 7
critical up-feed velocity 76
critical velocity 48, 100
cross-coupling 32
C type chips 141
cut-off burr 109
cutting-directional vibration-assisted (CDVA) 74
cutting-edge angle 89
cutting-edge radius 89
cutting energy 49, 69, 98
cutting force 8, 119–144
cutting force coefficient 127
cutting paths 132
cutting power 49
cutting temperature 105
cutting zone 89

Vibration Assisted Machining: Theory, Modelling and Applications,
First Edition. Lu Zheng, Wanqun Chen, and Dehong Huo.

d

data acquisition (DAQ) 66–67
data collection 66–67
decoupling 28, 45
DEFORM
  2D software 152
deformation zone 104
depth ratio 120
diffusion wear 97
drive and control 19
  open loop control system 19
  proportion integration differentiation (PID) 19, 66

e

effective cutting time 48
electrode plate 49
electromechanical coupling coefficient
elliptic vibration-assisted (EVA) cutting 74
energy-dispersive X-ray spectroscopy (EDS) 101, 105
excitation sources 47

f

ferrous materials 98
Fick's first law 105
finite element analysis (FEA) 17, 22, 63–65
fish scale surface 188
fish squamous structure 188
friction force 100
front cover 49–53

g

Gibbs free energy 104
Gibbs–Helmholtz equations 105

h

hard and brittle materials 1, 7, 36, 98, 102, 113
  AISI 1045 steel 90
  Steel and SiCp/Al composites 99
  titanium alloys 102
homogeneous transformation matrix 174
horizontal speed ratio (HSR) 76

i

ImageJ 172
independent drive 34–35
instantaneous uncut thickness 135
Inverse kinematic modeling (IKM) 56

j

Johnson–Cook damage model 89
Johnson–Cook (JC) material model 89

l

LabVIEW platform 66–67
Langevin transducer 50
linear Hooke's law 56

m

matrix-based compliance modeling (MCM) 56
maximum vibration velocity 85
mechanical amplification 18, 46
mechanical stiffness 29–30, 49
mechanical wear 96–98
micro-electro-mechanical systems 167
micro-Euler–Bernoulli beam 56
minimum chip thickness 106
motion strokes 33

n

natural frequency 17, 19, 22, 26, 28
negative shear angle 110
negative shear band 110
nodal plane 46, 52
node point 20, 27, 53
nominal cutting velocity 85
nonresonant system 27–36
normal-directional vibration-assisted (NDVA) 74
normal wear 95–96
  boundary wear 96
  flank wear 96
  rake wear 96
  tool tip wear 96

o

output compliance 61–62
oxidation wear 98

p

parasitic motions 33
piezoelectric transducer (PZT) 2
plastic-elastic deformation 106
plastic hinge point 110
ploughing effect 106
Poisson burr 109
polycrystalline diamond (PCD) tools 99
processing parameters 9, 169
  cutting depth 123
  feed rate 74, 83, 106
pseudo-rigid body (PRB) method 56
pulse force 47, 49

q

quarter-wave horns 53

r

radial force 136
rear cover 49–53
relative rotational angle 178
residual stress 108
resonance rod 27
resonant system 19–28
right circular flexure hinge 56–58
rollover burr 109
rotation matrix 60
rotation speed 87

s

scanning electron microscope (SEM) 99
secondary damage 107
shear angle 125
shear yield stress 110
signal generator 66–67
Sindatek Water Contact Goniometer 187
size effect 106
spindle rotation frequency 88
spindle rotation speed 185
stress concentration 22, 31, 64
structural compliance 63–64
structural stiffness 63–64
surface quality 9–10, 159
surface roughness 107, 108, 146
surface simulation 185
surface texture 10, 146, 167–171, 185–187
surface topography 188
sweep signal 66

t

tangential force 136
tear burr 109
thermochemical wear 96–98
thermo-elastic-plastic material 148
tool-chip contact length 152
tool geometry 108
tool holder 20–21, 30–31
tool life 9, 95–108
tool materials 35, 96, 145
tool parameters 9, 98, 170
tool paths 8, 87
  elliptical tool paths 87
tool tip motion 75
tool trajectory 25, 30, 101, 120, 155
  elliptical tool trajectory 101
tool-workpiece separation (TWS) 73–75, 80
top burr height 163
top burrs 112
transmission efficiency 47
transmission mechanisms 18–19
  acoustic waveguide booster 19–20
  double parallel four bar linkage 19, 32
  flexible hinges 18, 32, 33, 56–59
  hollow ultrasonic horn 22
  sonotrode 19–20
  ultrasonic horn 18–19, 22
  ultrasonic transducer 19–20

u

ultrasonic generator 46
ultrasonic transducer 46
ultrasonic vibration 22–25, 35
uncut work material 76
unloaded power 49–50
up-feed increment 76

v

vibration amplitude 18, 22, 26, 29, 49, 67, 78, 82
vibration assisted machining process 3

vibration assisted machining process (*contd.*)
  vibration-assisted drilling 3–4
  vibration-assisted grinding 4–6
  vibration-assisted machining process 3
  vibration-assisted milling 3
  vibration-assisted polishing 6
  vibration-assisted turning 4
vibration frequency 29, 36, 47, 49, 53–54, 74
vibration resolutions 33